AF790099

Philosophy of Engineering and Technology

Volume 50

Editors-in-Chief

Pieter E. Vermaas, Dept of Philosophy, Delft University of Technology, Delft, The Netherlands

Darryl Cressman, Dept of Philosophy, Maastricht University, Maastricht, The Netherlands

Editorial Board Members

Neelke Doorn, Dept of Philosophy, Delft University of Technology, Delft, The Netherlands

Edison Renato Silva, Federal University of Rio de Janeiro, Rio de Janeiro, Brazil

Philip Brey, University of Twente, Enschede, The Netherlands

Louis Bucciarelli, School of Engineering, Massachusetts Institute of Technology, Belmont, USA

Michael Davis, Humanities Dept, Illinois Institute of Technology, Chicago, USA

Andrew Feenberg, Simon Fraser University, Burnaby, Canada

Luciano Floridi, Dept of Philosophy, University of Hertfordshire, Hertfordshire, UK

Jun Fudano, Kanazawa Institute of Technology, Nonoichi, Japan

Sven Ove Hansson, Division of Philosophy, Royal Institute of Technology KTH, Stockholm, Sweden

Craig Hanks, Philosophy Dept, Texas State University, San Marcos, USA

Vincent F. Hendricks, Center for Information & Bubble Studies, University of Copenhagen, København, Denmark

Billy Vaughn Koen, Dept of Mechanical Engineering, University of Texas, Austin, USA

Peter Kroes, Dept of Philosophy, Delft University of Technology, Delft, The Netherlands

Sylvain Lavelle, Ctr for Ethics, Tech & Society, ICAM Paris-Senart Engineering School, Lieusaint-Senart, France

Michael Lynch, Cornell University, Ithaca, USA

Anthonie W. M. Meijers, Dept of Philosophy and Ethics, Eindhoven Univ of Technology, Eindhoven, The Netherlands

Duncan Michael, Ove Arup Foundation, London, UK

Carl Mitcham, Liberal Arts & International Studie, Colorado School of Mines, Golden, USA

Byron Newberry, Baylor University, Waco, USA

Helen Nissenbaum, New York University, New York, USA

Alfred Nordmann, Institut für Philosophie, Technische Universität Darmstadt, Darmstadt, Germany

Joseph C Pitt, Dept of Philosophy, Virginia Tech, Blacksburg, USA

Daniel Sarewitz, Consortium for Sci Policy & Outcome, Arizona State University, Washington DC, USA

Jon Alan Schmidt, Aviation & Federal Group, Burns & McDonnell, Kansas City, USA

Peter Simons, Trinity College Dublin, Dublin, Ireland

Jeroen van den Hoven, Dept of Philosophy, Delft University of Technology, Delft, The Netherlands

Ibo van der Poel, Dept of Philosophy, Delft University of Technology, Delft, The Netherlands

John Weckert, Centre for Applied Philosophy & Ethics, Charles Sturt University, Canberra, Australia

The *Philosophy of Engineering and Technology* book series provides the multifaceted and rapidly growing discipline of the philosophy of technology with a central overarching and integrative platform. It publishes on all topics in the philosophy of technology and is open to all research communities across the world.

Specifically, it publishes edited volumes and monographs in:

· the phenomenology, anthropology and socio-politics of technology and engineering

· the emergent fields of the ontology and epistemology of technology and design

· engineering ethics and the ethics of specific technologies ranging from nuclear technologies to artificial intelligence

· written from philosophical and practitioners' perspectives and authored by philosophers and practitioners

The series also welcomes proposals that bring these fields together or advance philosophy of engineering and technology in other integrative ways.

Proposals should include:

· A short synopsis of the work or the introduction chapter

· The proposed Table of Contents

· The CV of the lead author(s) or editor(s)

· If available: one sample chapter

We aim to make a first decision within 1 month of submission. In case of a positive first decision the work will be provisionally contracted; the final decision about publication will depend upon the result of anonymous peer review of the complete manuscript. We aim to have the complete work peer-reviewed within 3 months of submission.

The series discourages the submission of manuscripts that contain reprints of previous published material and/or manuscripts that are below 150 pages/75,000 words.

For inquiries and submission of proposals authors can contact either editor-in-chief: Pieter Vermaas at p.e.vermaas@tudelft.nl, or Darryl Cressman at darryl.cressman@maastrichtuniversity.nl

Alice Fox • Darryl Cressman
Editors

Arrangements of Power: Tracing Langdon Winner's Legacy Within and Beyond the Philosophy of Technology

Springer

Editors
Alice Fox
Old Dominion University
Norfolk, VA, USA

Darryl Cressman
Maastricht University
Maastricht, The Netherlands

Editors-in-Chief
Pieter E. Vermaas

ISSN 1879-7202 ISSN 1879-7210 (electronic)
Philosophy of Engineering and Technology
ISBN 978-3-032-17370-6 ISBN 978-3-032-17371-3 (eBook)
https://doi.org/10.1007/978-3-032-17371-3

© The Editor(s) (if applicable) and The Author(s), under exclusive license to Springer Nature Switzerland AG 2026

This work is subject to copyright. All rights are solely and exclusively licensed by the Publisher, whether the whole or part of the material is concerned, specifically the rights of translation, reprinting, reuse of illustrations, recitation, broadcasting, reproduction on microfilms or in any other physical way, and transmission or information storage and retrieval, electronic adaptation, computer software, or by similar or dissimilar methodology now known or hereafter developed.
The use of general descriptive names, registered names, trademarks, service marks, etc. in this publication does not imply, even in the absence of a specific statement, that such names are exempt from the relevant protective laws and regulations and therefore free for general use.
The publisher, the authors and the editors are safe to assume that the advice and information in this book are believed to be true and accurate at the date of publication. Neither the publisher nor the authors or the editors give a warranty, expressed or implied, with respect to the material contained herein or for any errors or omissions that may have been made. The publisher remains neutral with regard to jurisdictional claims in published maps and institutional affiliations.

This Springer imprint is published by the registered company Springer Nature Switzerland AG
The registered company address is: Gewerbestrasse 11, 6330 Cham, Switzerland

If disposing of this product, please recycle the paper.

To Langdon Winner and all of those who stand with him, shoulder to shoulder, to forge paths to futures arranged differently through systems designed to ensure things stay forever the same.

Preface

In the initial invitation for submissions, we intentionally sought work from a wide range of career stages, research interests, and geographies to try to encompass the far reach of Winner's work beyond the politics and philosophy of technology. To this end, we were successful in curating a collection of chapters that offer distinct and divergent perspectives on Winner's writing. One of the primary challenges of curating such a vibrant collection of chapters has remained an organizational one: none of the authors are working to answer a single question, rely on the same concept, or analyze the same technology. Thus, instead of a chapter-by-chapter overview or section categorization, we offer a few major throughlines that might help readers first entering the conversation more easily position each author relative to each other and the volume as a whole text.

The primary organizational structure of this volume is to draw the reader's attention to where we, as readers and thinkers-alongside, are positioned within politics. The chapters toward the front zoom out to provide a strong sense of "we are here" when it comes to approaching systems and structures of power through Winner's work. As the volume progresses, the perspective shifts to "you are with us" as the authors work to bring us into the fields, flows, and relationships through which they work. We hope this approach emphasizes the way in which these massive and daunting arrangements of power, like AI and democracy, move into the artifacts of the everyday and back again.

Although it was unintentional, Artificial Intelligence is one of the major throughlines of this volume. The initial invitation for chapters was released in the summer of 2023, right on the heels of major advancements in Generative AI by companies such as Open AI. At present, GenAI stands to be one of the most disruptive technologies in recent history, with major impacts on labor, industry, environment, and social trust already taking hold at the time of writing this introduction. As such, these chapters also offer an early snapshot of how folks thinking alongside Winner were reacting to GenAI technologies: their concerns, predictions, and hopes for the future.

Another major throughline is a watchful eye on the political cadence across the globe. While critical scholars across a variety of subsets of philosophy and technology have been carefully documenting the traces of fascism interwoven throughout our socio-material systems for quite some time through the concepts of techno-fascism, elite capture, and the dark enlightenment, the last 2 years have marked a notable increased pace to this end. Although many of the authors in this volume do not reside in the USA, careful attention is paid throughout to the way technologies and systems ambivalently support and challenge consolidations of power, narrative control, and human experiences. The authors' connections between particular social orders and technological systems, like AI, paint a troubling picture of one possible future in which history and reality are manufactured in data centers to ensure that human labor and intellect are not inefficiently wasted on seeking truth, justice, or equality.

The final connection that we want to emphasize is the sense of deep care that these authors have across all parts of the systems they examine. Even when authors are critical, there is an undeniable sense of the value and gravity given to the systems themselves, but also for the wide variety of actors who shape and are shaped by these relationships. While each author may take a different approach to Winner's work and their unique chapter, Winner's careful attention to interconnecting the forest and the trees, in balance, manifests throughout. As a result, these chapters push the reader to be continually situated in the in-between space of the whole and its parts simultaneously so that even the subtle threads of everyday power are scrutinized.

A core part of Winner's legacy as a thinker and writer is that his perspectives on power, technology, and social order leave us *changed* as learners and knowers. If one did arrive at Winner's work conceptualizing arrangements of power as "the way things are" or political artifacts as "neutral tools" to be used, one can never sleep comfortably again after reading *The Whale and the Reactor* or "Do Artifacts Have Politics?" and wondering where that leaves you, as a human, in these systems. And there is something so awfully delightful about the way that disquiet becomes an integral part of how one experiences and exists in the world. We hope that the reader finds the authors in this volume as illuminating and thought provoking as we have. But, perhaps even more boldly, we hope that reading these chapters leaves you somehow changed, deeply curious, and moderately unsettled and unsatisfied with our current power configurations.

Norfolk, VA Alice Fox

Introduction

For many people, Langdon Winner was their entry point into the philosophy of technology (and STS). I do not think that it is a stretch to say that his essay "Do Artifacts have Politics?" (1980) is responsible for teaching more people about the philosophy of technology than any other book or essay. His essay, or more specifically the case of low-hanging overpasses, has stuck for decades. This is no small feat. Social theorists of technology have conducted innumerable case studies of technologies, but only a few have achieved the status of Winner's bridges. We are all, in some way, students of Langdon Winner.

Academic fame of this sort, though, can have unintended consequences. The constant attempts at refutation are one thing and are to be expected; indeed, one can only hope that their ideas garner enough attention to be critiqued by people they've never met! The more problematic aspect of academic fame is that it can limit an otherwise impressive range of research to one paper, or more specifically, one particular case study. Having a conversation about Winner's reading of Marx, for example, or his critique of technocracy, tends to be overshadowed by a return to bridges; the academic equivalent of being turned into a "one-hit wonder." Not a bad fate, to be sure, but as it is in most of these cases, reducing Winner's work to highway overpasses is like reducing the Beatles to "I Want to Hold your Hand."

This reference to pop music is a good starting point for thinking about the scope of Winner's work. Besides being involved in protests at UC Berkeley in the 1960s and participating in what is now known as the New Left, he also wrote about music for Rolling Stone magazine, writing about musicians like Captain Beefheart, Neil Young, and Miles Davis. Who knows, in another world he could have pursued a path of influence as a music critic. His book *Autonomous Technology: Technics-out-of-Control as a Theme in Political Thought* (1977) is one of the first philosophical critiques of technology written by an American, although the book was, and still is, sometimes categorized as a work of political science or political philosophy. He was one of the first philosophers to draw on Wittgenstein to think about technology (Winner, 1983), he wrote about appropriate technology and the Office of Technology Assessment (OTA) as important social movements by which citizens can have greater influence over the trajectory and aims of technological change, and he was an active member of the Society for Philosophy and Technology since its early days. He has also not been shy about his place within the history of the philosophy of technology and saw his own work as continuing the work of critical philosophers of technology who preceded him, including Karl Marx, Rachel Carson, Jacques Ellul, and Herbert Marcuse. In this way, reading Winner for many was also an introduction to breadth of ideas and writers that make up the discipline of the philosophy of technology.

Across his work, one constant has been Winner's enduring sense of humor in the face of the absurdities of modern technological society. Humor among philosophers is rare but important; a way to critique that can ward off fatalism or despair. To take one example, he developed a parody of industry and academic boosters of technologies intended to revolutionize education. Winner adopted the persona of L.C. Winner, CEO of Educational Smart Hardware Alma Mater, Inc. (EDU-SHAM) who worked his way through a slick slide show presenting the Automatic Professor Machine. For anyone who has ever listened to stories of how MOOCs, VR goggles, or AI assistants will forever change education, remembering Winner's Automatic Professor Machine provided needed comic relief.

Shameless tech-promotionalism is only one of the objects of Winner's varied philosophical critique. In his early work, it is a political critique, aimed at technocracy, technological determinism, the ideology of progress, and the ways that technologies are used to create long-lasting power imbalances that are difficult to undo. His example of how nuclear power may infringe upon civil liberties is a good example of how technologies can make unpleasant politics inevitable and nearly impossible to change. As he points out, choosing nuclear power means also choosing a state that requires a military apparatus to safeguard the uranium and plutonium used in nuclear power plants. This inherently political technology can legitimate kinds of reasoning "that justify the adaptation of social life to technical requirements" (1986 [1980], p. 38). Going further, we could also point out that radioactive waste requires storage and although there are solutions that would enable the storage of waste for tens of thousands, if not millions of years, the question remains—who, or what, will manage the safety of this waste for such a period of time? It is unlikely that states, as we know them today, and their militaries, will exist as they do for 10,000 years. Technology, in this way, is a form of politics that extends across geographical space and over very long periods of time.

This is a good example of how Winner does not shy away from talking about the hard problems confronting technological modernity that, unfortunately, are largely missing from contemporary philosophical discussions of technology. If Winner's work was so influential, why has a generation of philosophers of technology forgotten the uncompromising critique that he conveyed through his work? For example, I found this quote amongst the writings of a leading contemporary philosopher of technology who encourages readers of the futility of opposing technology. Using the case of Facebook, he writes, "Facebook is simply one of the media through which friendship can take place…rather than alienating people from each other, Facebook mediates their relations, offering new ways of interaction" (Verbeek, 2013, p. 78). Thinking of a smiling Mark Zuckerberg and other software barons cozying up to Donald Trump should be enough to have us running back to Winner for lessons in how to think about technology. But this is only one example of what is a strand of conservative politics running through contemporary philosophy of technology that discourages any form of opposition, resistance, and rejection of technology in favor of developing tools for living with technology: "we should not negate our current capitalist society and its concurrent technologies, but seek for manners to develop a good life amidst these relations of power" (Rao et al., 2015, p. 459). The power of resisting technology, of saying "no" and rejecting the technologies that are on offer from corporations no longer seems like a legitimate position amongst philosophers. In a recent interview with Alfred Nordmann, Winner laments this situation, describing much work in the philosophy of technology as follows:

> The attitude is to support and carefully monitor the fascinating technoscience developments in the making. As the processes of Research & Development unfold and reach culmination a scholar may find opportunity to comment upon interesting properties in the workings of the various devices, systems, algorithms, and other novelties, identifying their interesting ethnographic, philosophical and ethical features. But the basic understanding is one fully characteristic of twentieth century technothink: Innovate first. Ponder the implications later. In that light there is a consistent disposition to encourage potentially world changing developments to unfold and to offer erudite, retrospective (but likely irrelevant) commentaries as the fascinating prospects emerge. The possibility that an ethically or politically autonomous human presence might intervene in time to make a difference is seldom if ever on the agenda. The idea that one might announce a firm "no" to any attractive, innovative pathway is simply out of the question (Winner, 2020).

This sort of ego deflation is part of Winner's charm (if one does not mind being deflated). In his essay "Brandy, Cigars, and Human Values" (1986), he takes apart the discipline's obsession with "values." This talk of values is largely empty, he argues, "a hollow discourse that…usurps much of the space formerly occupied by much richer, more expressive categories of moral and political language" (p. 156). Winner's solution? "In many instances, the word embarrassment would serve just as well" (p. 162). Imagine, journals would be renamed as "Science, Technology, and Human Embarrassments" while research programs would be known as "Embarrassment-Sensitive Design." We could do the same for other terms that have neutered critique, like "responsible," "disruptive," or "smart." Funny, to be sure, but also a jolt to researchers and writers who are too eager to align themselves with the ambitions of industry and the state to prioritize values, responsibility, or ethics; but only after, and not in opposition to, growth, productivity, and profit.

This points to a bigger question: what place should Langdon Winner play in today's philosophy of technology? As the papers in the collection can attest to, the politics of technology is still a fruitful path of philosophical inquiry, albeit not what it once was. Philosophical critique of technology has not run out of steam, rather it has been marginalized and pushed to the periphery among a long list of less problematic and more polite approaches that avoid implicating designers, policymakers, venture capitalists, engineers, and investors as implicit, and at times explicit, proponents of deeply problematic politics intended to disadvantage people and harm the environment. Perhaps then, this is Winner's enduring lesson, beyond bridges: the arrangements of power that are materialized by technology are deeply unfair and biased to favor the interests of those who fund, build, and administer technologies. It is the task of philosophers to draw attention to the processes by which these power imbalances occur, no matter who is offended.

Maastricht University
Maastricht, The Netherlands
Darryl Cressman

References

Rao, M., Jongerden, J., Lemmens, P., & Ruivenkamp, G. (2015). Technological mediation and power: postphenomenology, critical theory, and autonomist Marxism. *Philosophy and Technology, 28*, 449–474.

Verbeek, P. P. (2013). Resistance is futile: Toward a non-modern democratization of technology. *Techne: Research in Philosophy and Technology, 17*(1), 72–92.

Winner, L. (1977). *Autonomous technology: Technics-out-of-control as a theme in political thought*. MIT Press.

Winner, L. (1980). Do artefacts have politics? *Daedalus, 109*(1), 121–136.

Winner, L. (1983). Technologies as forms of life. In R. S. Cohen & M. W. Wartofsky (Eds.), *Epistemology, methodology and the social sciences*. Kluwer Academic.

Winner, L. (1986). Brandy, cigars, and human values. In *The whale and the reactor* (pp. 155–164). University of Chicago Press.

Winner, L. (2020). Langdon Winner, autonomous technology—Then and still now. Interview with Alfred Nordmann. *Jahrbuch Technikphilosophie,* 295–301.

Contents

Chapter 1
(How) Do Artefacts Have Politics? (Re-)Tracing the Ontic and the Ontological in Winner's Legacy

Anthony Longo and Mark Coeckelbergh

Abstract This chapter revisits Langdon Winner's question "Do Artefacts Have Politics?" to argue that contemporary philosophy of technology has overly confined political analysis of technologies to an ontic register, focusing on discrete artefacts and their immediate social effects. We reconstruct an ontological approach that investigates how technologies help constitute the very conditions of possibility of politics. First, drawing on Heidegger's ontic/ontological distinction, we show how the empirical turn reads Winner's 1980 essay in isolation and thereby reinforces an ontic focus. Second, we retrieve the neglected ontological impulses in Winner's later account of "technologies as forms of life" supplemented with the transcendental "grammars" that pre-shape technological meaning and use. Third, we connect this line of thought to Arendt's analysis of artefacts as the material "in-between" that enables political intersubjectivity, re-reading Winner's Moses' bridge case as world-structuring rather than merely behavior-shaping. Finally, we extend this framework to social media algorithms, arguing that their processual, constantly recalibrating character inaugurates a distinct, indeterminate way of "having politics" that cannot be captured by the classic politics-of-artefacts template. The chapter thus proposes a productive entwinement of ontic and ontological inquiry for analysing political technologies in the algorithmic age.

Keywords Ontic-ontological distinction · Forms of life · Wittgenstein · Hannah Arendt · Politics of algorithms

---

A. Longo (✉)
Department of Philosophy, Tilburg University, Tilburg, The Netherlands
e-mail: a.n.longo@tilburguniversity.edu

M. Coeckelbergh
University of Vienna, Vienna, Austria
e-mail: mark.coeckelbergh@univie.ac.at

© The Author(s), under exclusive license to Springer Nature Switzerland AG 2026

A. Fox, D. Cressman (eds.), *Arrangements of Power: Tracing Langdon Winner's Legacy Within and Beyond the Philosophy of Technology*, Philosophy of Engineering and Technology 50, https://doi.org/10.1007/978-3-032-17371-3_1

1.1 Introduction

With his article "Do Artefacts Have Politics?" (1980), Winner set the course for the contemporary study of the relationship between politics and technology. His thinking gained traction in the context of arguments against general analyses of Technology-at-large in phenomenology and critical theory and in favor of the concrete study of human-technology relations in everyday life. It thus became part of the empirical turn (Achterhuis, 2001) and postphenomenology (e.g. Ihde, 1990). However, because of this influence, political theory of technology has mainly concerned itself with the 'ontic' level of description, focusing on the direct social histories and political consequences of artefacts. This is especially problematic if we want to account for the politics of algorithmic technologies such as AI and blockchain, that typically move beyond the one-on-one relation between an individual and an artifact, and which influence is interwoven in complex ways with the social and political realm.

In this chapter, we argue that this approach risks losing sight of those aspects of technologies and their influence that cannot be reduced to an ontic level but require an 'ontological' analysis into the conditions of possibility for politics that artefacts also create. While the dominance of ontic-level inquiry over ontological analysis may have been influenced by Winner's power-based approach, we claim that this is more so due to a reading of his 1980-article in isolation from his other work that does address ontological concerns. Therefore, we propose an ontological approach to the politics of artefacts that escapes these worries. As a starting point for such an approach, we return to Winner's ([1983] 2001) often overlooked application of Wittgenstein's concept of 'forms of life' to technology. We further unpack this approach drawing from Coeckelbergh's 'grammatical' approach to technology and connecting this to Arendt's ontological approach to politics, in which artefacts provide the transcendental ground for political intersubjectivities and power relations.

The chapter proceeds in five steps. In Sect. 1.2, we clarify the distinction between 'ontic' and 'ontological' with the help of Heidegger's critique of Western metaphysics and show how current thinking about the politics of artefacts, as influenced by Winner, fits into the 'ontic' level of analysis. Section 1.3 then returns to Winner's (2001) article "Technologies as Forms of Life" as a starting point for ontological inquiry into technology to make clear what is lost when overemphasizing the ontic level of being at the cost of the ontological. This loss is compensated by introducing Coeckelbergh's (2022) 'grammatical' approach to technology that considers the conditions of possibility that already pre-shape (ontologically) what we do with technology before the actual (ontic-level) interaction with technology. In Sect. 1.4, we use a subtle reference to Arendt in a footnote by Winner (2001) to connect Winner's Wittgensteinian approach to technology with Arendt's ontological approach to politics. In Sects. 1.5 and 1.6, we use Arendt's framework to elaborate on Winner's ontological concerns, first theoretically and then more concretely, by going back to Winner's case of Moses' bridges and moving forward with an analysis of social media algorithms. Throughout this article, we suggest ways in which ontic and ontological levels of inquiry can each be put to use for a more productive and nuanced analysis of the politics of artefacts.

1.2 The Ontic and the Ontological in Philosophy of Technology

So-called 'classical philosophy of technology' gets its name, as well as its controversies, from its specific way of analyzing technology. Mainly inspired by Martin Heidegger, this tradition is explicitly concerned with an 'ontological' inquiry into technology as a mode of Being. Heidegger famously argued in *The Question Concerning Technology* ([1954] 1977) that technology is more than just an instrument to help people achieve certain goals. It embodies, rather, a certain way of understanding the world that characterizes a historical era. Modern technology, for example, structures our relation to the world in such a way that everything appears as a 'standing reserve', always-available for us to exploit it. In criticizing modern technology, Heidegger is not concerned with concrete artefacts and their properties but with Technology as a mode of revealing the world. Speaking in his own terms, we must move from an 'ontic' level of inquiry—focusing on the concrete properties, behaviors and arrangements of entities, i.e. whatever can be known *empirically* about artefacts—toward an 'ontological' analysis, asking what it is for a certain class of entities (e.g. artefacts) 'to be' in the world. Heidegger warns that if we skip this ontological inquiry into the different 'ways of being' of animals, artefacts, human beings, etc., we risk flattening our understanding of various kinds of entities in the world based on mere surface resemblances (Slaby, 2021, 542). However, as Heidegger writes, 'these ontological foundations can never be disclosed by subsequent hypotheses derived from empirical material', rather 'they are always "there" already' (BT 75/50). The task of the philosopher, then, is to make explicit these ontological foundations that are always-already there and how they structure the concrete behaviors we can study empirically. In other words, classical philosophy of technology proceeds in the style of a 'transcendental analysis' into the conditions of possibility of concrete human-technology relations rather than the technologies themselves (Verbeek, 2005, 7).

This transcendental approach is precisely the main target of the 'empirical turn' (Achterhuis, 2001). While Heidegger saw it as crucial to move 'upwards' toward ontological analysis, scholars like Peter-Paul Verbeek and Andrew Feenberg were very critical of *imposing* ontological frames 'top-down' onto the ontic level of existence. The way philosophers like Heidegger but also Horkheimer and Adorno ([1947] 2016), Ellul ([1964] 2011) and Marcuse ([1964] 2003) tackled philosophical questions about technology was deemed insufficient because of its generalizing and monolithic conception of technology. The source of this tendency, according to Feenberg, is that they allow ontological presuppositions to 'intrude unacceptably on the ontic level', covering up the different ways that people can engage with technology in specific social and historical contexts (Feenberg, 2010, 449). Moreover, the neglect of the *actual* relations that people enter into with technology (on ontic level) results in an 'aporetic' stance that leaves us hanging with a conception of technology as a totally binding force yet still allowing exceptions, making modern technology into an 'incurable disease with a cure' (Belu & Feenberg, 2010). While classical positions in the field tend to reduce technological artefacts, systems and practices to their ontological conditions, the empirical turn urged philosophers of technology to take the empirical reality—or as Heidegger would say: artefacts in their 'ontic concretion' (Heidegger, 2007, 78)—as a point of departure to bring to light the social and political effects of particular technologies (Verbeek, 2022, 39).

In this context, scholars have regularly turned to Winner's seminal article "Do Artefacts Have Politics?" (1980) to illustrate what the empirical turn aims to uncover (e.g. Achterhuis, 2001; Ihde, 2010, 82; Smits, 2001; Verbeek, 2022, 40). His analysis of the political properties of artefacts like Robert Moses' bridges in New York and McCornick's molding machines gained traction precisely because of its focus on the direct, ontic effects of concrete technologies. Although Winner does not explicitly engage with the ontic-ontological distinction, the 'ontic' level is clearly prioritized over the ontological question. Anticipating on later critiques of social constructivism in STS, Winner notes that a political analysis of technology is incomplete if it does not account for the concrete properties of the technologies at stake: 'Rather than insist that we immediately reduce everything to the interplay of social forces, the theory of technological politics suggests that we pay attention to the characteristics of technical objects and the meaning of those characteristics' (Winner, 1980, 123). At the same time, merely a description of these characteristics will not suffice either; in being more-than-instrumental, technologies encompass 'purposes far beyond their immediate use' (Winner, 1980, 125). We therefore always also need to understand the technical, societal and political arrangements that *precede* the use of the artefacts in question (Winner, 1980, 125).

Technologies become *inherently* political when these man-made systems *require* particular kinds of political relationships:

> the adoption of a given technical system actually *requires* the creation and maintenance of a particular set of social conditions as the operating environment of that particular system. [...] In this conception, some kinds of technology require their social environments to be structured in a particular way in much the same sense that an automobile requires wheels in order to run. The thing could not exist as an effective operating entity unless certain social as well as material conditions were met (Winner, 1980, 130).

Elsewhere, he writes that the tangible effects of technological innovation are the 'result of a new system's own *operating requirements*: it simply will not work unless human behavior changes to suit its form and processes' (Winner, 1993, 11). One could read these passages as a 'transcendental' analysis, emphasizing the conditions of possibility of certain technologies that have to be met in order for these technologies to be or become political. However, Winner continues immediately by correcting this interpretation: 'The meaning of "required" here is that of practical (rather than logical) necessity' (Winner, 1980, 130). This helps us understand the 'ontic' demarcation of Winner's approach in this article. In line with his later critique of social constructivism (cf. Winner, 1993) he already proposed in 1980 a move from social *properties* that define the context of making and using technology towards the societal *arrangements* that correspond with, or are *required by*, certain technologies. In doing so, Winner constantly flirts with the transcendental but never goes so far. This flirting comes back when he distinguishes 'internal' conditions inseparable from 'the workings of a given technical system' and 'external' conditions 'removed from the organization of those technologies as such' (Winner, 1980, 130).

Many scholars within STS, political philosophy and philosophy of technology have followed Winner's plea to consider the political properties of artefacts by examining their technical properties as well as their corresponding socio-political contexts. Postphenomenologists have built on Winner's power-based approach by analyzing the kinds of human-technology relations that enable these power relations in the first place by shaping people's perceptual and agentic capabilities (Verbeek, 2020, 143). Robert Rosenberger, in particular, has demonstrated at length in *Callous Objects* (2017) how technical mediations can be used to discriminate against specific groups (e.g. benches with armrests making it impossible for unhoused people to lay down). Lucas Introna and David Wood have applied Winner's framework to show how digital systems, like facial recognition software, are political in the same sense as Moses' bridges: each, 'by its very design, includes certain interests and excludes others'. (Introna & Wood, 2004, 179). Again in comparison with Moses' bridges, Bucher argues that algorithms are a 'really blatant example of that same logic as they explicitly decide, order, and filter the world in specific ways' (Bucher, 2018, 23).

Some have taken Winner's approach as a baseline to expand his framework. Jenny Lee Davis, for instance, explicitly positions her work in *How Artefacts Afford: The Power and Politics of Everyday Things* as an extension of Winner's 'classic work' to think of technologies as 'encoded with power relations that produce patterned effects' (Davis, 2020, 55). While Winner sticks to what technologies may 'require', Davis distinguishes different ways in which artefacts can participate in power relations: artefacts may 'demand' or 'refuse' people to act in certain ways (in a strong sense), but they may also 'request', 'encourage', 'discourage' or 'allow' particular behavior. These ways of affording may overlap and depend on who it applies to: for instance, low hanging bridges may 'allow' small vehicles to pass under, while 'refusing' buses to do so. Davis calls these mechanisms 'analytic stopping points that help describe the intensity with which technological objects facilitate or impede particular lines of action and social dynamics' (Davis, 2020, 65).

Note that these 'conditions' still remain within the ontic level of inquiry. Both the technical and political arrangements, as well as the internal and external conditions of technologies that make them possible, are 'objective' states of affairs that we can empirically describe from an third-person perspective. In Heideggerian terms, the objects of study (i.e. the artefacts, the societal context, the required political relationships) are equally treated as 'present-at-hand' (*Vorhanden*) within a 'theoretical attitude', and not—as an ontological inquiry demands—as involved in the world in their respective ways of Being. Therefore, if we follow Heidegger, the empirical approach evades the 'real *philosophical* problem' (Heidegger, 1977, 45), namely, to understand how the interaction between artefacts, societal arrangements and political relationships is possible in the first place because of their specific way of Being.

These shortcomings of the empirical turn have motivated certain scholars to call for a rehabilitation of the 'ontological' in philosophy of technology (e.g. Coeckelbergh, 2017, 2022; Lemmens, 2022; Lemmens et al., 2017; Misa, 2009; Zwier et al., 2016). However, it seems that, so far, debates within philosophy of technology are mostly split into two opposite 'camps', with the empirical-ontic approach on the one side, and the transcendental-ontological approach on the other. However, this split need not be as dichotomous as is often presented. As Nolen Gertz's (2018) study on nihilism and technology, for instance, illustrates, it is possible to use empirical reality as a starting point to build up towards transcendental analysis. Indeed, as Heidegger himself continually warned in *Being and Time*, the move from ontic to ontological is unsuccessful if it is a move *away* from the ontic. Although the distinction between the two remains philosophically crucial for him, the ontic level is the 'stepping stone' necessary for reaching the ontological in the first place (Slaby, 2021, 542). This is a fundamental insight for philosophy of technology that has not been given full consideration yet. As a thorough read of Heidegger's work shows, 'the relationship between the ontic encounter and the a priori ontological structure of being-with cannot be characterized as a simple priority of the latter over the former' (McMullin, 2013, 89–90). The *difficulty* of philosophical inquiry is precisely to keep distinct that which is, in practice, always entwined: concrete ontic encounters *evoke* the ontological categories or *existentialia* through which they are understood, and at the same time we must account for the very possibility to *receive* these evocations.

1.3 Retracing the Ontological in Winner

Although at first sight Winner's approach entails a prioritization of the ontic level of analysis, we interpret his overall body of work in a way that reveals a much more nuanced and complex appreciation of both the ontic and ontological levels of technological mediation. A more contextual reading of his separate works indicates that his turn to the ontic level in "Do Artefacts Have Politics?" does not necessarily imply a complete abandonment of ontological concerns. This is exemplified by his ambiguous relation with Ellul's 'classical' approach in *The Technological Society* (2011), as 'Winner's feelings towards Ellul oscillate between love and hate' (Smits, 2001, 155). In fact, Winner himself warned throughout his works against completely giving up what classical philosophy of technology has taught us: 'Before we forget our Marx or our Mumford, Ellul, or Heidegger, it is important to notice what one gives up as well as what one gains in choosing the intellectual path to the study of technology and human affairs' (Winner, 1993, 367). A purely empirical-descriptive approach has certain advantages, though they come at 'significant cost' (Winner, 1993, 368). Even if social constructivism succeeds in opening the 'black box' of different social actors and relationships that socio-technical arrangements often are, 'the box they reveal is still a remarkably hollow one' (Winner, 1993, 375). What they lack, according to Winner, is not only an evaluative stance towards certain technical systems but also a fundamental philosophical reflection on 'the larger question about technology and the human condition', i.e. how technologies may affect the 'basic commitments and projects' of our current age (Winner, 1993, 375). So, while Winner, in line with the empirical turn, takes issue with Heidegger's and Ellul's generalizing and monolithic conceptions of technology, he also does not go so far to completely disregard ontological analysis.

Still, given that his engagement with the ontological is mostly only *implicit*, the next section will turn to Arendt's critical appropriation of Heideggerian phenomenology to fully draw out the ontological dimensions in Winner's overall philosophy of technology. Before we do that, however, we need to understand how 'the ontological' enters Winner's philosophy of technology.

In an often overlooked paper, "Technologies as Forms of Life" ([1983] 2001), Winner uses Wittgenstein's term 'forms of life' to argue that we should not only study the "impact" of technologies but also how technologies restructure our physical and social worlds and, ultimately, our lives. Just as Wittgenstein argued that the speaking of a language is not just about *using* words but a way of shaping our practices, Winner suggests that in adopting technologies, 'they become woven into the texture of our everyday existence' and eventually 'shed their tool-like qualities to become part of our very humanity' (Winner, 2001, 12). In other words, the instrumentality of technologies itself leads to its 'more-than-instrumental' character, since it is precisely through their use, in everyday contexts, that technologies shape humanity:

> We do indeed "use" telephones, automobiles, electric lights, and computers in the conventional sense of picking them up and putting them down. But our world soon becomes one in which telephony, automobility, electric lighting, and computing are forms of life in the most powerful sense: life would scarcely be thinkable without them (Winner, 2001, 11).

In this sense, technology is a strong driver in shaping cultural practices. But as Winner rightly argues, the influence goes in both directions. On the one hand, technologies bring about 'significant alterations in patterns of human activity and human institutions'—they make 'new worlds' (Winner, 2001, 11). Yet, on the other hand, the forms of life in which new technologies enter and are embedded also shape our use, interactions, and expectations. As Wittgenstein wrote in his "Philosophy of Psychology" (published with the *Investigations*), that forms of life are the 'given,' 'what has to be accepted' (Wittgenstein, [1954] 2009). This 'given' exerts its influence on the use of technology and related interactions. In a similar spirit, Winner writes that 'most of the transformations that occur in the wake of technological innovation are actually variations of very old patterns', giving the example of the humanization of computers:

> Forms of life that we mastered before the coming of the computer shape our expectations as we begin to use the instrument. One strategy of software design, therefore, tries to "humanize" the computers by having them say "Hello" when the user logs in or having

> them respond with witty remarks when a person makes an error. We carry with us highly structured anticipations about entities that appear to participate, if only minimally, in forms of life and associated language games that are parts of human culture (Winner, 2001, 14).

Technologies-in-use are interwoven and interdependent on social and cultural patterns and roles that existed before the innovation took place—patterns that are altered in the wake of the innovation, when the technologies are used. With Winner, we can place technologies in the context of the 'vast multiplicity of cultural practices that compromise our common world' (Winner, 2001, 15).

In this regard, our use of technologies are deeply embedded in what Wittgenstein calls 'language games', again in two directions. Winner's example of the computer, cited above, illustrates very well the relationship between technology and language. Existing linguistic frameworks continuously pre-mediate both our making and using of technologies. Metaphors often play a key role in how we make sense of new technologies and integrate them in our daily lives, both making visible *and* hiding certain aspects of technologies: 'each machine metaphor has opened areas of both insight and radical blindness as it becomes means of interpreting what happens in our world' (Winner, 1977, 45). As he notes in his critical study of the concept of 'autonomous technology', the continuing changes between the machines that shape our social and political relationships, and the metaphors that we use to make sense of them explain how technology is a permanent 'source of concern' (Winner, 1977, 45). Yet because of this, technologies also disrupt the very language we use ourselves to make sense of society and of ourselves. We cannot always count on existing words and metaphors to grasp the meaning of new technical and political arrangements. This struggle is obvious in how we currently are trying to make sense of AI-driven technologies like ChatGPT: should we call them autonomous writers, assistants, stochastic parrots, or something entirely different? How we speak of technologies, what we name them, which verbs and adjectives we associate with them, all illustrate that 'language reflects the content of technical practice'. And what Wittgenstein can teach us here, Winner suggests, is that 'we would expect that changes of this kind would appear, sooner or later, in the language people use to talk about themselves' (Winner, 2001, 15).

Using vocabulary from phenomenology one could say that Winner is suggesting here that technology has an 'ontological' dimension and not only an 'ontic' one: politics of technology is not only about 'what things do' (to use Verbeek's famous phrase), that is, their more-than-instrumental impact here and now in and through a particular use and interaction, but *also* about what structures technology and what gives it meaning. Compared to "Do Artefacts Have Politics?" (1980), here Winner already starts moving 'upwards' to a transcendental level of inquiry, moving his concern from mere 'practical requirements' (Winner, 1980, 130) toward the 'logical' requirements that enable technology to become political in the first place. Yet to further develop this argument we have to move beyond Winner. While later Winner returns to Wittgenstein when exploring the connection between language games and technology (Winner, 2018), acknowledging Coeckelbergh's work but mainly focusing on language in the sense of the rhetorics about technology, he never developed the more structuralist and ontological approach to (the politics of) technology itself he suggested in the "Forms of Life" paper. This can be done, however, by drawing on Coeckelbergh's "grammatical" approach (Coeckelbergh, 2018, 2022).

Coeckelbergh crafted his approach in order to move beyond the phenomenology and hermeneutics of individual human-technology relations as found in postphenomenology (Ihde, 1990; Verbeek, 2005), which focuses on individual artefacts without taking into account the role of language and (other) more structural elements that also shape the meaning and influence of technology. Taking inspiration from Wittgenstein, Ricoeur, Bourdieu, and Cassirer, Coeckelbergh proposes a structuralist and transcendental approach to what technology is and means. His use of the term 'transcendental' does not make technology into a large Heidegerrian abstraction, as Ihde worried, but is rather mobilized to claim that 'the use and meaning of technology are made possible by, and structured by, some other elements that are not themselves necessarily technological and material. These structures or conditions pre-shape the meaning of, and our performances with, technology.' (Coeckelbergh, 2018, 151). In order to develop this conceptually, he expands on Winner's adoption of the term 'form of life' from the later Wittgenstein (2009). He argues that the role of the transcendental structure of technology is analogous to that of the transcendental structure of language: like language use, technology's use is also embedded in a form of life, in a broader structure that gives it meaning. Technology is part of our social and political world and must be understood within its "grammars". At the same time, both language use and technology use actively shape that wider world. Grammar is given and rather resilient, but not entirely. Structures move. This opens up room for change.

Coeckelbergh (2022) identifies four structures that expand on Winner's framework and function as transcendental conditions or 'grammars' of technology: language, social relations, human bodies, and material infrastructures. He gives the example of Artificial Intelligence (AI): what AI is and means depends on how we speak about it and thus literally on grammar (for example when we address a particular chatbot as "she" this taps into the structures of our language) and on the narratives that circulate in our culture, but also, for example, on family relations and gender relations: the way we speak to an AI assistant may be shaped by structural gender biases in our society. Language games already shape how we use and interact with certain technologies, even *before* we actually engage with them. This interaction also has a particular meaning and impact, of course, but one should not neglect the grammars of use that already pre-shape that very meaning and interaction. This also has normative implications. When we evaluate technology, we also have to evaluate its grammars, its transcendental conditions 'the challenge of understanding and questioning technology means nothing less than the challenge of understanding and questioning our ways of being in the world—indeed our form of life' (Coeckelbergh, 2022, 159). In that sense, the proposed phenomenology and hermeneutics is political. Questioning technology becomes also questioning the social order.

This is in line with the spirit of Winner's work at large and with the initial paper on 'forms of life'. While, as Winner later explains, he comes more from the political theory side rather than, say, phenomenology, his use of the term opens up interesting avenues of doing politics of technology if interpreted in the light of Coeckelbergh's theory. It shows that next to paying attention to the use of words about technology (rhetorics about technology), which is needed since language is part of the grammar of technology, Winner's project of a politics of technological artefacts can be expanded to a broader transcendental approach to technology that considers a range of (other) background conditions that 'structure' technology use, and in that way does fully justice to the *critical* potential in Winner's work. Starting from Winner's suggestion that we must attend to forms of life, we have thus not only retracted the ontological in Winner but also further developed what this means. In the next sections, we continue this move from the ontic to the ontological in response to Winner drawing from Arendt's critical engagement with the ontological difference in her analysis of modern technology.

1.4 Re-Reading Winner with and Along Arendt

The choice to mobilize Arendt's analysis of modernity to expand on Winner's reflections on technology is less than obvious. Not only is Arendt rarely taken seriously as a philosopher of technology (with a few exceptions, e.g. Longo, 2024; Markell, 2011; Undurraga, 2019), in even fewer occasions is her work connected to Winner's. This strikes us as a missed opportunity, given the regular references one finds across Winner's texts to several aspects of Arendt's political theory. For Winner, Arendt is one of those few 'radical thinkers' (Winner, 2018, 301) who had thoroughly examined the repercussions of modern technology for 'what it means to be human and what constitutes the "human condition"' (Winner, 2001, 56). If, as Winner claims, 'The basic task for a philosophy of technology is to critically examine the nature and significance of artificial aids to human activity' (2001, 4), there is no doubt that Arendt is a philosopher of technology. Moreover, wherever Winner cites Arendt, he mostly does so to suggest *potential* elaborations on his own analyses without developing them further himself. This section therefore will unpack Winner's suggestions in an Arendtian spirit.

Although both come from different traditions, they often reach similar conclusions about the relationship between politics and technology. In these moments, they both continually rely on the two-way influence between technology and language. While technology is seen as disruptive, it can only disrupt our lives because it enters into the pre-understanding (*Vorverständnis*) of the common world. Winner's claim that technologically induced transformations occur as 'variations of very old patterns' corresponds with Arendt's hermeneutic orientation in which encountering the world always happens by drawing from long standing interpretative frameworks (Borren, 2010, 23). It is in this context that both testified of the impression that the atom bomb had had on them (Arendt, [1958] 2018, 6; Winner, 1980, 131): the first atomic explosion had set into motion a 'process' of which the consequences could no longer remain in anyone's hands. Such inherently political technologies were not just new 'instruments' for those in power, they became part of the new conditions under which political life from now on would unfold (Arendt, 2018, 9): 'once artefacts such as nuclear power plants have been built and put in operation, the kinds of reasoning that justify the adaptation of social life into technical requirements pop up as spontaneously as flowers in the spring' (Winner, 1980, 134). While originally citizens acting together in the public realm had the privilege of setting into motion such processes (which, for Arendt, is what the capacity for action is all about), modern society has made scientists the true actors of our time (Arendt, 2018, 323; Winner, 1977, 96).

For these reasons, Arendt and Winner were very critical of the notion—if not the *metaphor*—of 'progress' to speak of politics and technology. Thinking history in terms of process rather than a series of related but contingent events was a result of the rising dominance of the so-called 'natural sciences' in the nineteenth century, in which the biological 'life process' as a set of causal sequences became the model to understand all human affairs, including political and technical practice (Arendt, 2006, 61). This did not remain a mere intellectual problematic but one with practical effects, Winner argues:

> Why has a culture so firmly based upon countless sophisticated instruments, techniques, and systems remained so steadfast in its reluctance to examine its own foundations? Much of the answer can be found in the astonishing hold the idea of "progress" has exercised on social thought during the industrial age. In the twentieth century it is usually taken for granted that the only reliable sources for improving the human condition stem from new machines, techniques, and chemicals (Winner, 2001, 5).

By the same token, both rejected any attempt to 'predict' the effects of political and technical actions on the common world. For Arendt, these attempts in which 'permanence is entrusted to a flowing process' (Arendt, 2006, 75) goes back to Plato, 'the first to design a blueprint for the making of political bodies' and who's approach 'has remained the inspiration of all later Utopias' (Arendt, 2018, 227). Although the figure of the 'philosopher-king' is rarely defended anymore in contemporary debates, 'mainstream philosophy is still caught in the same trap [...] making redundant the haphazard contingency of accommodations reached in actual political arenas' (Canovan in Arendt, 2018, xviii). Winner similarly notes that such urges to predict which changes technology will bring about—'a somewhat more farsighted version of technology assessment' (Winner, 1980, 10)—is still very visible in debates on the social impact of digital technologies.

The unease with modes of predicting the effects of technological developments was indeed one of the main issues that Arendt and Winner had with Heidegger. While Winner more specifically has taken issue with Heidegger's monolithic and over-imposing ontological conception of 'Technology with capital T', Arendt, for similar reasons, challenged Heidegger's ontological project in its entirety. Instead of conceptualizing human existence in terms of *existentialia*, Arendt spoke of the pregiven and man-made *conditions* of human existence (Arendt, 2018, 7). This implies a shift in attention from 'modes of Being' to 'modes of activity' (cf. *Vita Activa*, the title of the German version of *The Human Condition*). Attending to historical shifts in the composition of such activities, and the 'conditional structures' they enact, allows her to theorize paradigmatic shifts without relying on a 'history of Being' as Heidegger proposes. By framing every occurrence within such ontological frames, Heidegger is left with 'a concept of the Self as the total opposite of man' (Arendt, 2005, 181), devoid of all human characteristics that 'arise from human spontaneity' (Arendt, 2005, 178), or what she later called 'natality' (Arendt, 2018, 9). Still, Arendt's project does not therefore also reject any ontological reflection. As Martin Braun remarks, there is more to Arendt's analysis than ontic descriptions of 'what we are doing' (Arendt, 2018, 5): she 'goes back to the ontological conditions through which these activities are made possible in the first place' (Braun, 1994, 19).

Inserting this contingency as the basic condition of the relationship between politics and technology resulted for both in a more-than-instrumental conception of technology. As we already suggested in the previous section, Winner accounted for the non-neutrality of technologies by arguing how they correspond with certain 'forms of life'. Arendt had already taken issue with the instrumentalist and anthropocentric views of technology that Winner was arguing against:

> The discussion of the whole problem of technology, that is of the transformation and world through the introduction of the machine, has been strangely led astray through an all-too-exclusive concentration upon the service or disservice the machines render to men. The assumption here is that every tool and implement is primarily designed to make human life easier and human labor less painful. Their instrumentality is understood exclusively in this anthropocentric sense (Arendt, 2018, 151).

Later on, Winner similarly noted that '[j]udgments about technology have been made on narrow grounds, paying attention to such matters as whether a new device serves a particular need, performs more efficiently than its predecessor, makes a profit, or provides a convenient service' (Winner, 2001, 11). Their responses are alike: where Winner writes that the world-making capacity is 'the most important accomplishment of any new technology' (Winner, 2001, 11), Arendt agrees that 'homo faber, the toolmaker, invented tools and implements in order to erect a world, not—at least, not primarily—to help the human life process' (Arendt, 2018, 151). Therefore, any major technological disruption results in '[v]ast transformations in the structure of our common world' (Winner, 2001, 9).

Rereading Winner through an Arendtian lens helps to see how Winner continuously and productively moves between ontic and ontological ways of thinking. In putting in the center of philosophy of technology the question 'what kind of world we are making' whenever we 'make things work', noting how the everyday use of tools inevitably correlates with the 'production of psychological, social, and political conditions as a part of any significant technological change' (Winner, 2001, 17), ontic descriptions of what we are doing leads us automatically to the ontological task of analyzing how technologies become the transcendental conditions of our very existence, as well as how they themselves rely on other conditions like language, social relations, human bodies, and material infrastructure (Coeckelbergh, 2022).

1.5 How Artefacts Had/Have Politics: Changes in the Human Condition

By now it has become clear that Winner's overall concern in his philosophy of technology, in the legacy of Arendt, was the 'interesting puzzle [...] that we so willingly sleepwalk through the process of reconstituting the conditions of human existence' (Winner, 1980, 10). One of the major ways that technology plays a role in this is by not only changing our existing practices and 'forms of life', but by *adding* 'fundamentally new activities to the range of things human beings do' (Winner, 2001, 13). He mentions in this regard the changes that 'work' (as *poiesis*) is undergoing in modern society. As Marx taught us, being human is being 'homo faber': production is not just one of many activities that we can or cannot engage in, it is part of the fundamental ways of expressing life. And if Winner and Arendt jointly agree with Marx on one aspect, it is his claim that modern technologies have not only 'aided' how we do 'work' but also powerfully redefined what 'work' means (Winner, 2001, 6). Elsewhere, in an often-unnoticed paper, Winner (1969) has drawn attention to the political effects of cybernetics—a theme that Arendt (1964) also only addressed as an explicit theme in one conference paper but whose more fundamental concerns with the larger transformations of which cybernetics are only part were extensively discussed in *The Human Condition*. On these occasions, Winner refers to Arendt for further analysis, though did not seem to realize that a more elaborate reflection on this matter also has consequences for thinking the politics of artefacts as he did in 1980. This one but last section then wants to show that a more thorough awareness of the repercussions of Arendt's analysis might have given Winner reasons to expand his own framework of political technologies.

Arendt offers throughout *The Human Condition* a critical take on the changing relationship between politics and technology in the modern age. Her analysis rests on a particular appreciation of artefacts and their role in the political realm that so far has been underappreciated in Arendt scholarship (cf. Coeckelbergh, 2009; Longo, 2024). This neglect is mostly due to an *immaterialist reading* of her political theory, in which 'action' is ontologically separated from 'work' and 'labor', the two activities that are explicitly concerned with the material environment. However, this reading ignores how the 'web of human relationships', as Arendt calls it, is only formed through a constant intersubjective relation to the material world of objects or '*Dingwelt*': 'To live together in the world means essentially that a world of things is between those who have it in common' (Arendt, 2018, 52). This material environment acts as an 'objective in-between' that allows people to share words *about* some 'worldly objective reality,' something that literally is between people (*inter-est*) (Arendt, 2018, 182). The products of homo faber then, as we saw, are not just tools to serve human ends; they erect a world by enabling and structuring intersubjective encounters: 'as a table is located between those who sit around it; the world, like every in-between, relates and separates men at the same time' (Arendt, 2018, 52). As a result, there is a special role for homo faber in making the actions of *zoon politikon* in the public realm possible (or not).

As Ihde and Malafouris have pointed out, today we are witnessing a transformation of homo faber itself, as new 'modes of making' the *Umwelt* arise due to the advent of smart technology: we become embedded in 'active technological environments' (Ihde & Malafouris., 2019, 322). Although they embed the discussion in the recent technological developments, Arendt already early on identified these changing conditions with the rise of automation and cybernetics: 'the present technology has changed the very worldliness of the human artifice' (Arendt, 2018, 150). What 'technology' *is* changes along with the world it erects. It no longer appears as a stable objective in-between, but increasingly 'like a biological development of mankind in which the innate structures of the human organism are transplanted in an ever-increasing measure into the environment of man' (Heisenberg quoted in Arendt, 2018, 153). Arendt takes this transformation as a reason the re-evaluate the *vita activa* itself. In addition to labor, work and action—the basic activities with which we express human existence on earth—needs to be supplemented with a new activity.

Where Arendt calls this new activity in the English version of *The Human Condition* a form of 'creating' (*machen*), the German version adds the term '*technisch tun*' ([1981] 2020, 175). For Arendt this has caused a rupture in the activity of 'work' itself, since the classical categories to describe the products of homo faber no longer seem to apply. While traditionally, fabrication involved a 'series of separate steps' in which the product was still clearly distinct from the process of making, technology has become 'a continuous process' that dissolves this very distinction (Arendt, 2018, 149). Again, this transformation does not just entail a change in 'what we are doing' (ontically) but has ontological implications as well: rather than conceiving of events, things and individuals in their singularity, this 'process turn' re-frames phenomena as (part of) autopoietic, self-organizing and all-encompassing processes (Hyvönen, 2016, 14). It would be shortsighted to conclude that the advent of automation has simply *distorted* human perception; rather, this process-framing became an intrinsic *part of* how we apprehend the common world, 'bridging the gap between human subjectivity and nonhuman processes' (Simbirski, 2016, 608). Arendt and Winner come together smoothly here: technologies do not just affect changes in the activities we engage in but also—perhaps more importantly—in the everyday conditions under which we engage in these activities, leading to transformations of the 'basic commitments and projects of modern technological society' (Winner, 1993, 375).

1.6 Back to Moses' Bridge, and Moving on with Algorithms

Confronting this analysis with Winner's framework suggests that there may be different *ways in which* artefacts can have politics. Here, again, the importance of a fruitful engagement with both the ontic and the ontological level of technological mediation becomes clear, as well as the role that Arendt can play in continuing

Winner's and Coeckelbergh's move from empirical to transcendental philosophy of technology.

To make this concrete, we revisit Winner's classic case of Moses' bridge from an ontological perspective. In Winner's ontic account of the politics of artefacts, Moses' bridges are political because they practically correspond with or 'require' certain political relationships (e.g. structural and racist patterns of equality). An ontological reading of this example, however, shows the bridges do more than just directly enabling and constraining certain actions for certain groups of people (or installing 'affordances', as Davis (2020) calls it). They also erect a 'world' in which the racist segregation is maintained and becomes part of a 'way of life'. As Coeckelbergh (2022) argued, such examples only be accounted for if we *assume* other conditions that make these influences possible. For example, the capacity to make a distinction between 'black' and 'white' rests on the fact that human existence is mediated by *language*. Moreover, the way human existence is ontologically always already *embodied* as well as embedded in *social relationships* with others are conditions of possibility for such 'ways of life' to emerge in the first place. Lastly, artefacts 'make worlds' because of their creation of and relation to *material and infrastructural environments* which we commonly relate to, as Arendt notes. In doing so, artefacts that make up public space (such as Moses' bridges) become an 'objective in-between'.

Yet, it is important to see that artefacts *can* gather, relate and separate us in certain ways only *because* their way of existing (or 'ontological constitution') enjoys stability and permanence. By transcending the cyclical temporality of nature and life processes, artefacts endow the human world with durability (Arendt, 2018, 137; Taminiaux, 1997, 14). Bonnie Honig has expressed the importance of such objects in her book *Public Things: Democracy in Disrepair* (2017). At this level man-made things start to provide a so-called 'holding environment'. Stable artefacts allow political communities to cohere by extending a kind of magical envelope around us that forms people as a collective. Yet, importantly, they do so without determining in any way what a collective means or does. Good examples of this are public parks or other public meeting spaces. Such objects often provide the framework for shared practices that support a sense of community among people. This enables artefacts—as "in-between"—to become a matter of shared concern: '[m]ost action and speech is concerned with this in-between' (Arendt, 2018, 182). It allows us to talk about their meaning, as Winner himself exemplifies. Yet, at the same time, the robustness of material artefacts also enables them to discriminate consistently in line with certain political relationships, as Winner's examples illustrate. If Moses' bridges would not enjoy a durable and stable existence, as *thingly* in-between, it would not succeed in 'having politics' the way Winner describes.

It is therefore no exaggeration to say that social media algorithms are deeply transforming the structure of political relationships: the ontological status of algorithms fundamentally undermines the thingly stability that characterizes traditional artefacts like Moses' bridges. While material artefacts maintain a durable presence that persistently structures human movement and social access, the contemporary algorithmic infrastructures of social media are marked by a processual character that resists fixity. This distinction is particularly evident in the shift from *static*, rule-based algorithms (also 'symbolic AI')—designed to consistently produce the same output for a given input—to *dynamic* machine learning systems that generate models from data and continuously recalibrate their behavior (also 'sub-symbolic AI') (Bucher, 2018, 24; Benjamin, 2023, 14). Their purpose is to find probabilistic models that can approximate patterns in the world (Benjamin, 2023, p. 223). Rather than executing a prescribed function or a set of rules, the purpose of ML is to *find*—perhaps even 'diagnose'—a function to make sense of certain input.

Arendt already anticipated this shift in 1958, noting that modern technologies were undergoing as well as amplifying a 'process turn', in which the 'product' of homo faber can no longer be properly distinguished from the 'process' of making it (Arendt, 2018, 150). As Andreas Hepp has pointed out, digital media are not simply 'there' but always *emerging* in ongoing interactions between hardware, software and human activity: 'based on algorithms and infrastructures, they are generated in narrow recursiveness loops and exist in a state of continuous "beta"' (Hepp, 2020, 12). Therefore, it is nonsensical to say that we would 'see' or 'use' an algorithm (like we would see and use an artefact); enact the algorithm through human-computer interaction (Galloway, 2006, p. 19). Machine Learning (ML) algorithms are never pre-set but emerge as a 'strategy, or plan of action—based on interactions with unfolding events' (Chun, 2013, p. 126); any social media algorithm thus constantly evolves along with the data it draws from. This process of 'finding' predictive functions is not only an activity of 'the algorithm' itself. Also, the user is involved in helping the algorithm find a set of rules that fits their needs or desires. Its continuous 'improvement' is not a linear development based on automated inputs that eliminate human involvement but is always a *temporary* improvement serving a specific user-context. Algorithms are therefore better understood as 'performative events' dependent on the situatedness of their actions and responding to particular circumstances rather than stable and universal rules. Therefore, because algorithms have a different 'way of existing' compared to artefacts or things, the 'forms of life' that we engage in are also shaped by our 'processual' apprehension of individuals and events in the world. And this, in turn, causes algorithmic technologies to exemplify a different way of 'having politics': This means that algorithmic systems—whether filtering news feeds, curating content, or structuring visibility—do not simply embed pre-defined political logics into durable infrastructures. Rather, they enact an ongoing process of reconfigurating the contours of political agency through an evolving interaction between datafied inputs, user responses, and platform affordances. In such an environment, mediation itself is no longer a function of stable technological forms but a continuous negotiation between human intentionality and the predictive operations of AI systems.

This does not mean that social media environments lack stability entirely. As scholars within the growing field of platform studies have shown, platforms do maintain structural possibilities and constraints through, for instance, moderation policies and affordances that shape user behavior. Platforms such as Twitter (now: X) and TikTok establish socio-technical infrastructures that enable new ways of constructing collective narratives and identities (Longo, 2023). Decentralized publishing, the interactivity of the medium, and the persistence and replicability of online content (Boyd, 2011, 46) all provide a level of infrastructural durability akin to traditional artefacts. For these reasons, Lucas Introna and David Wood (2004) have argued that digital technologies, such as facial recognition software, replicate the logic of Winner's analysis by embedding social and political biases into their design: each, 'by its very design, includes certain interests and excludes others' (Introna & Wood, 2004, 179). Bucher (2018) extends this analysis by explicitly likening social media algorithms to Winner's case of Moses' bridges: just as Moses' bridges physically regulated access to public infrastructure along racial and class lines, algorithms today 'explicitly decide, order, and filter the world in specific ways' (Bucher, 2018, 23).

However, we claim that the *transformative* political effects of such platforms arise less from these durable platform configurations and more from the processual character of social media algorithms as 'in-between'. That is, the instability and indeterminacy of algorithmic environments have become defining characteristics of contemporary social media platforms themselves, shaping how content appears and how users respond to it. Platforms like TikTok, Instagram, and YouTube operate through continuously evolving algorithmic infrastructures that transform user engagement into input for further recalibration. This means that what is 'common' in algorithmic environments—both in the sense of what gathers, relates, and separates users and in the sense of what is collectively experienced as a matter of concern—is not a stable, thingly in-between, but an indeterminate, *processual in-between*. Jurgita Imbrasaite (2022) similarly argues that TikTok's algorithmic infrastructure fosters an *emergent* public realm in which users 'act-with' the platform's algorithm, rather than merely 'acting in' a pre-existing environment. This is particularly evident in TikTok's For You Page (FYP), where users do not simply engage with a static social space but are continuously re-situated within a personalized, evolving media environment that reconfigures itself in response to their behaviors: 'Rather than being a pre-defined technological structure that determines political action, the algorithm emerges as an infrastructure that is itself shaped and reshaped through user interaction' (Imbrasaite, 2022, 15). This indeterminacy is then the new 'quasi-transcendental' condition under which political interaction in algorithmic environments come into being. Since we are still, as both scholars and as society-at-large, trying to grapple with the consequences of these changes, finding a language to adequately articulate this new mode of 'having politics' will remain a challenge for the coming years, if not decades. This preliminary analysis shows that much more research can be done within the framework that Winner and Arendt have helped us set up.

Taking Winner's approach to the politics of artefacts into the digital age means also seeing this ontological distinction. We cannot just project the structure of the 'politics of artefacts' onto the 'politics of algorithms'. Moreover, if we stick to a strict separation between ontic and ontological concerns with technology, we risk conflating different ways in which technologies *can be* political. As Heidegger had always intended it, and as Winner and Arendt support it, philosophers of technology now face the task of moving beyond the ontic/ontological *dichotomy* into a productive *entwinement* between the two. The exploration of how artefacts have politics in this chapter reveals a nuanced relationship between the empirical level of human-technology relations and the transcendental conditions that they enact. By extending Winner's framework, informed by Arendt's analysis, we see that artefacts do more than merely influence, enable and constrain behavior; they help construct the very worlds we inhabit, shaping the structures of social and political life. As technologies evolve, particularly with the rise of algorithmic processes, the way these worlds are formed and maintained changes along and therefore fundamentally alters how politics is embedded within technological configurations. As we move further into the digital age, recognizing and philosophically analyzing these ontological shifts will be crucial for comprehending the new political relationships that Winner has managed to make us see.

References

Achterhuis, H. (2001). *American philosophy of technology: The empirical turn* (R. P. Crease, Trans.). Indiana University Press.

Arendt, H. (1964). *Hannah Arendt papers: Speeches and writings file, 1923-1975; Essays and lectures; "Cybernetics," lecture, 1964*. Manuscript/Mixed Material. MSS11056, Box 72. Library of Congress. https://www.loc.gov/item/mss1105601187/

Arendt, H. (2005). What is existential philosophy? In J. Kohn (Ed.), *Essays in understanding, 1930-1954: formation, exile, and totalitarianism* (pp. 163–187). Schocken Books.

Arendt, H. (2006). *Between past and future: eight exercises in political thought* (Penguin classics). Penguin Books.

Arendt, H. (2018). *The human condition*. The University of Chicago Press.

Arendt, H. (2020). *Vita activa oder Vom tätigen Leben*. T. Meyer (Ed.). Piper.

Belu, D. S., & Feenberg, A. (2010). Heidegger's aporetic ontology of technology. *Inquiry, 53*(1), 1–19. https://doi.org/10.1080/00201740903478376

Benjamin, J. J. (2023). *Machine horizons: Post-phenomenological AI studies*. PhD, University of Twente. https://doi.org/10.3990/1.9789036555357

Borren, M. (2010). *Amor Mundi: Hannah Arendt's political phenomenology of world*. PhD, University of Amsterdam. http://dare.uva.nl/record/346972

Boyd, D. (2011). Social network sites as networked publics: Affordances, dynamics, and implications. In Z. Papacharissi (Ed.), *A networked self: Identity, community and culture on social network sites* (pp. 39–58). Routledge.

Braun, M. (1994). *Hannah Arendts tranzendentaler Tätigkeitsbegriff*. Lang.

Bucher, T. (2018). *If...then: algorithmic power and politics*. Oxford University Press.

Chun, W. H. K. (2013). *Programmed visions: Software and memory*. MIT Press.

Coeckelbergh, M. (2009). The public thing: On the idea of a politics of artefacts. *Techné: Research in Philosophy and Technology, 13*(3), 175–181.

Coeckelbergh, M. (2017). *Using words and things: language and philosophy of technology*. Routledge.

Coeckelbergh, M. (2018). Technology games: Using Wittgenstein for understanding and evaluating technology. *Science and Engineering Ethics, 24*(5), 1503–1519. https://doi.org/10.1007/s11948-017-9953-8

Coeckelbergh, M. (2022). The grammars of AI: Towards a structuralist and transcendental hermeneutics of digital technologies. *Technology and Language, 3*(2), 148–161. https://doi.org/10.48417/TECHNOLANG.2022.02.09

Davis, J. L. (2020). *How artefacts afford: the power and politics of everyday things*. MIT Press.
Ellul, J. (2011). *The technological society: A penetrating analysis of our technical civilization and of the effect of an increasingly standardized culture on the future of man* (J. Wilkinson, Trans.). Vintage Books.
Feenberg, A. (2010). *Between reason and experience: essays in technology and modernity*. MIT Press.
Galloway, A. R. (2006). *Gaming: Essays on algorithmic culture*. University of Minnesota Press.
Gertz, N. (2018). *Nihilism and technology*. Rowman & Littlefield International.
Heidegger, M. (1977). *The question concerning technology, and other essays* (W. Lovitt, Trans.). Garland Publishing.
Heidegger, M. (2007). *Being and time* (J. Macquarrie & E. Robinson, Trans.). Blackwell.
Hepp, A. (2020). *Deep mediatization. Key ideas in media and cultural studies*. Routledge.
Honig, B. (2017). *Public things: democracy in disrepair*. Fordham University Press.
Horkheimer, M., & Adorno, T. W. (2016). *Dialectic of enlightenment* (J. Cumming, Trans.). Verso.
Hyvönen, A.-E. (2016, November). Invisible streams: Process-thinking in Arendt. *European Journal of Social Theory, 19*(4), 538–555. https://doi.org/10.1177/1368431016633572
Ihde, D. (1990). *Technology and the lifeworld: from garden to earth* (The Indiana series in the philosophy of technology). Indiana University Press.
Ihde, D. (2010). Heidegger's Technologies: Postphenomenological Perspectives. Fordham University Press.
Ihde, D., & Malafouris, L. (2019, June). Homo faber revisited: Postphenomenology and material engagement theory. *Philosophy & Technology, 32*(2), 195–214. https://doi.org/10.1007/s13347-018-0321-7
Imbrasaite, J. (2022). Acting-with: On the development of a public realm on TikTok during the pandemic and its potential to enable action. *Techné: Research in Philosophy and Technology, 26*(3), 504–522. https://doi.org/10.5840/techne2023126171
Introna, L., & Wood, D. (2004). Picturing algorithmic surveillance: The politics of facial recognition systems. *Surveillance & Society, 2*(2/3). https://doi.org/10.24908/ss.v2i2/3.3373
Lemmens, P. (2022, December). Technologizing the transcendental, not discarding it. *Foundations of Science, 27*(4), 1307–1315. https://doi.org/10.1007/s10699-020-09742-5
Lemmens, P., Blok, V., & Zwier, J. (2017). Toward a terrestrial turn in philosophy of technology. *Techné: Research in Philosophy and Technology, 21*(2/3), 114–126. https://doi.org/10.5840/techne2017212/363
Longo, A. (2023). Digital reconfigurations of collective identity on Twitter in advance: A narrative approach. *Techné: Research in Philosophy and Technology, 27*(1), 60–85. https://doi.org/10.5840/techne2023317175
Longo, A. (2024). Reversing the primacy of political action: Thinking politics and technology with Arendt. *Arendt Studies*. https://doi.org/10.5840/arendtstudies20248759. Online first.
Marcuse, H. (2003). *One-dimensional man: Studies in the ideology of advanced industrial society*. Routledge.
Markell, P. (2011). Arendt's work: On the architecture of the human condition. *College Literature, 38*(1), 15–44. https://doi.org/10.1353/lit.2011.a409860
McMullin, I. (2013). *Time and the shared world: Heidegger on social relations*. Northwestern University Press.
Misa, T. J. (2009). Findings follow framings: Navigating the empirical turn. *Synthese, 168*(3), 357–375. https://doi.org/10.1007/s11229-008-9447-y
Rosenberger, R. (2017). *Callous objects: Designs against the homeless*. Forerunners. University of Minnesota Press.
Simbirski, B. (2016). Cybernetic muse: Hannah Arendt on automation, 1951–1958. *Journal of the History of Ideas, 77*(4), 589–613. https://doi.org/10.1353/jhi.2016.0032
Slaby, J. (2021). Ontic (*Ontisch*). In M. A. Wrathall (Ed.), *The Cambridge Heidegger Lexicon* (pp. 542–546). Cambridge University Press. https://doi.org/10.1017/9780511843778.147

Smits, M. (2001). Langdon Winner: Technology as a shadow constitution. In H. Achterhuis (Ed.), *American philosophy of technology: The empirical turn* (R. P. Crease, Trans.) (pp. 147–170). Indiana University Press.

Taminiaux, J. (1997). *The Thracian maid and the professional thinker: Arendt and Heidegger* (SUNY series in contemporary continental philosophy. M. Gendre, Trans.). State University of New York Press.

Undurraga, B. (2019). Historicizing distinctions: Hannah Arendt on science and technology. *Arendt Studies, 3*, 153–172. https://doi.org/10.5840/arendtstudies201941518

Verbeek, P.-P. (2005). *What things do: Philosophical reflections on technology, agency, and design* (R. P. Crease, Trans.). Pennsylvania State University Press.

Verbeek, P.-P. (2020). Politicizing postphenomenology. In G. Miller & A. Shew (Eds.), *Reimagining philosophy and technology, reinventing Ihde* (Philosophy of engineering and technology) (Vol. 33, pp. 141–155). Springer. https://doi.org/10.1007/978-3-030-35967-6_9

Verbeek, P.-P. (2022). The empirical turn. In S. Vallor (Ed.), *The Oxford handbook of philosophy of technology* (pp. 34–54). Oxford University Press. https://doi.org/10.1093/oxfordhb/9780190851187.013.4

Winner, L. (1969). Cybernetics and political language. *Berkeley Journal of Sociology, 14*, 1–17.

Winner, L. (1977). *Autonomous technology: Technics-out-of-control as a theme in political thought*. 9. printing. MIT Press.

Winner, L. (1980). Do artefacts have politics? *Daedalus, 109*(1), 121–136.

Winner, L. (1993). Upon opening the black box and finding it empty: Social constructivism and the philosophy of technology. *Science, Technology, & Human Values, 18*(3), 362–378. https://doi.org/10.1177/016224399301800306

Winner, L. (2001). *The whale and the reactor: A search for limits in an age of high technology*. University of Chicago Press.

Winner, L. (2018). Technological investigations: Wittgenstein's liberating presence. *Techné: Research in Philosophy and Technology, 22*(3), 296–313. https://doi.org/10.5840/techne2018111485

Wittgenstein, L. (2009). *Philosophische Untersuchungen/Philosophical investigations* (G. E. M. Anscombe, P. M. S. Hacker, & J. Schulte, Trans.). Wiley-Blackwell.

Zwier, J., Blok, V., & Lemmens, P. (2016). Phenomenology and the empirical turn: A phenomenological analysis of postphenomenology. *Philosophy & Technology, 29*(4), 313–333. https://doi.org/10.1007/s13347-016-0221-7

Chapter 2
Artificial Intentions: Winner's Inherently Political Artifacts, AI and Government

Esther Keymolen and Donovan van der Haak

Abstract In his renowned paper *Do Artifacts Have Politics?*, Langdon Winner (*Daedalus, 109*(1), 121–136, 1980) ascribes two political qualities to technological artifacts. Firstly, Winner shows that technological artifacts can be used to settle social or political problems, and can thereby structure social relations, reinforce vested interests, and engineer human relations (i.e., politicized artifacts). Winner also discusses a second political quality of artifacts, arguing that some technological artifacts practically require or strongly attract a specific set of social and political conditions to operate (i.e., inherently political artifacts). We apply the insights from Winner's theory on the politics of artifacts to the use of artificial intelligence (AI) within government and argue that Winner's thought-provoking analysis of inherently political artifacts deserves more attention. Analysing how AI is and can be politicized helps us to see what social or political problems AI can assist governments in solving and how it may do so with important public values in mind (e.g., value-sensitive design). Dissecting the inherent politics of AI enriches our understanding of how AI's need for data, expertise, and infrastructure, requires a new, horizontally-orientated governance structure, while also attracting public-private partnerships and pre-sorting certain environmental and labor conditions. Recognizing the inherently political character of AI is necessary to be able to resist the increasing power of the private sector in government. This article provides new insights into how the socio-political arrangements that artifacts attract may threaten key public values and entrench existing power inequalities.

Keywords Artificial intentions · Politicized artifacts · Political artifacts · AI governance · Winner

---

E. Keymolen · D. van der Haak (✉)
Tilburg Institute for Law, Technology & Society (TILT), Tilburg University, Tilburg, Noord-Brabant, The Netherlands
e-mail: e.l.o.keymolen@tilburguniversity.edu; d.vdrhaak@tilburguniversity.edu

© The Author(s), under exclusive license to Springer Nature Switzerland AG 2026

A. Fox, D. Cressman (eds.), *Arrangements of Power: Tracing Langdon Winner's Legacy Within and Beyond the Philosophy of Technology*, Philosophy of Engineering and Technology 50, https://doi.org/10.1007/978-3-032-17371-3_2

2.1 Introduction

In the domain of the philosophy of technology, there is currently considerable attention to the normative questions concerning disruptive technologies such as artificial intelligence (AI) and related data-driven products and services. What are fair AI practices? How can we ensure that AI applications adhere to public values? What does human dignity come down to in a data-driven society?

To answer these and other normative questions, scholars predominantly turn to ethics, eliciting even new branches of ethics such as data ethics and AI ethics (Keymolen & Taylor, 2021). With a focus on data and algorithms on the one hand, and engineering practices on the other, it is the clear aim of these new ethics approaches to cover both the technological and human aspects of developing and using AI (Floridi & Taddeo, 2016). Moral theories and methods, from virtue ethics (Vallor, 2016) to thought experiments like the trolley problem (for a discussion of the trolley dilemma see: Nyholm & Smids, 2016), are revisited to analyze the normative aspects of AI and come up with solutions such as value-driven design strategies (La Fors et al., 2019; Umbrello & Van de Poel, 2021) and new ethical codes of practice (Boddington, 2017).

However, in the normative evaluation of AI, insights from political philosophy are not as widespread, barring exceptions, such as Coeckelbergh (2022), Van Maanen (2023), Sattarov (2019) and Saetra (2020). A gap seems to exist between the philosophy of technology and political philosophy (Coeckelbergh, 2022, p. 4). This is unfortunate, as many of today's technological challenges clearly would benefit from an approach that brings to the fore their political and systemic facets.

In the 1980s and 1990s, political philosopher, Langdon Winner made a convincing case for acknowledging and exploring the political qualities of technologies and understanding technology in a Wittgensteinian fashion as a "form of life", rather than merely a mouldable instrument in the hands of social actors (Winner, 1986). It is not so much that Winner does not agree that people, willingly and unwillingly, shape technologies. On the contrary, he affirms that technology is a crucial strategy for organizing society and that all sorts of—implicit—beliefs and values are incorporated in and eventually enacted by the technologies that societies run on. When governments[1] develop and use technologies they should take responsibility for it, for instance by firmly investing in technology assessments at the start of a project when options are still available (Winner, 2021).

This view also underpins most current regulatory, philosophical, and ethical analyses of AI, resulting in technological solutions such as "privacy-by-design" strategies (Gürses et al., 2011; Hoepman, 2014; Van der Hof & Lievens, 2018) and

[1] For the purpose of this article, we define governments broadly as any public institution or system with the authority to govern a country or state. We will not distinguish between local, national or supranational governmental institutions, as we will analyse AI's inherent politics applicable to different levels of government by dissecting the governmental use of AI on three levels: data, infrastructure, and expertise.

explainable AI (XAI) software (Adadi & Berrada, 2018; Ploug & Holm, 2020). Also, in the political and regulatory realm, there is a strong belief in responsible innovation through formulating stringent requirements for the development and use of AI. The European Commission declared that AI must be trustworthy, enhance human capabilities and uphold democratic principles. Practitioners in the field of AI are tasked with addressing ethical dilemmas and trade-offs "in a rational and methodological manner" (AI-HLEG, 2019, p. 20). They should "translate" legal and ethical requirements for trustworthy and human-centric AI into practical procedures that can be integrated into the design and application of AI systems (Keymolen, 2024, p. 14).

Less attention is being paid to the steering force of AI itself, what Winner refers to as inherently political qualities. Whereas it is widely acknowledged that some social actors are better placed than others to enact their political powers during a technological trajectory, the idea that some technologies are more resistant than others to these adjustments and rather bring along their own demands is not often part of the conversation. The overarching argument of this chapter is that these inherent political qualities, particularly when it comes to AI in government, deserve to be at the forefront of our attention.

Section 2.2 will succinctly set out Winner's key ideas on the politics of artifacts. Next, we will provide an overview of the arrival and use of AI in government (Sect. 2.3). We then use Winner's politics of artifacts to analyse the political qualities of AI within government (Sects. 2.4 and 2.5) which is followed by the discussion (Sect. 2.6) and conclusion.

2.2 Winner's Politics of Artifacts[2]

A fundamental premise in Winner's work (1980) is that artifacts are not neutral; they are carriers of political power. Implicitly and explicitly, the values and presuppositions of designers inevitably permeate their designs, thereby influencing the utilization of the artifacts (politicized artifacts). The capacity to embed political views into artifacts is a potent ability in a society where technology is essential for its functioning. Winner posits that artifacts can also have political ramifications in and of themselves, fostering and even demanding specific power relations (inherently political artifacts). He examines several instances where technological artifacts mediate practices to analyze such political effects.

Politicized Artifacts

In "Do artifacts have politics?", Winner transitions from discussing the intended political functions of artifacts to their unintended political impacts. The most frequently cited example of an intentional political artifact is Robert Moses' overpasses. Here, Winner describes how Robert Moses, the city planner responsible for

[2] Based on Keymolen (2016, pp. 135–137).

the design of highway overpasses leading to Jones Beach in New York, designed these overpasses so that only cars could pass under. The low-hanging overpasses prevented buses, which predominantly served poor African-American citizens, from accessing the beach. Thus, Moses' design embodied his racist vision, encompassing "purposes far beyond their immediate use" (Winner, 1980, p. 125).

This example sparked considerable debate among scholars (Joerges, 1999a, 1999b; Woolgar & Cooper, 1999) who challenged the analysis, suggesting that the overpass was not the sole route to Jones Beach or arguing that Moses was no more racist than his contemporaries (also see: Verbeek, 2011: 44).

However, the influence of artifacts is not confined to deliberate inscriptions by their developers that guide specific practices or, as with Moses' overpass, exclude certain groups. Winner also illustrates how public spaces, unintentionally designed with able-bodied individuals in mind, exclude differently-abled people. Similarly, he analyzes the introduction of an automated tomato harvester, which disadvantaged small farmers and radically transformed the agricultural sector. Designers can invent artifacts that, once integrated into society, provoke unforeseen and undesirable changes.

Inherently Political Artifacts

Subsequently, Winner discusses inherently political technologies. These are artifacts that, like the tomato harvester, generate unintended political consequences but differ in that they lack feasible design alternatives. The overpass was not required to be low-hanging, nor did public spaces cease to be public when designed for accessibility. Alternative designs can perform their intended functions without these particular negative consequences. Conversely, technologies such as nuclear plants or laboratories handling dangerous chemicals lack this flexibility. Choosing such technologies implies "choosing unalterably a particular form of political life" (Winner, 1980, p. 128).

Winner differentiates between a strong and a weak version of this claim regarding inherently political artifacts. The strong version asserts that "the adoption of a given technical system actually requires the creation and maintenance of a particular set of social conditions as the operating environment of that system" (Winner, 1980, p. 130). The weak version suggests that "a given kind of technology is strongly compatible with, but does not strictly require, social and political relationships of a particular stripe" (ibid). Both versions may apply to an artifact, indicating that while some aspects are immutable for the artifact to function, other aspects, though leaning towards a certain setting, can still be modified.

Winner notes that determining what constitutes a "practical necessity" and is therefore immutable is not merely an empirical question. He argues that the assertion 'it cannot be done any other way' often unjustly overrides other arguments. He writes (Winner, 1980, p. 133):

> …to say that some technologies are inherently political is to say that certain widely accepted reasons of practical necessity—especially the need to maintain crucial technological systems as smoothly working entities—have tended to eclipse other sorts of moral and political reasoning.

Winner does not adopt a deterministic stance, contrary to many classical philosophers of technology. Instead, he opposes their view, as determinism hinders accountability for the political consequences of artifacts. Rather, he illustrates that the openness or "multistability" (Ihde, 1990) is not uniform across all artifacts. Some artifacts are more amenable to modification by users or society than others. Consequently, the decision to introduce a particular technology is critical, as the potential for future adjustments is limited. Winner's analysis critiques an overly optimistic belief in the social power to shape artifacts' meaning and use. Not all artifacts are equally sensitive to social actors' influence, and not all social actors can shape artifacts' integration into daily life. As Winner (2001, p. 15) observes:

> The smart people are those able to 're-engineer' their organizations and careers by liquidating older roles, relationships, and institutions in response to technical and economical necessities that loom ahead. The less proactive to these conditions are doomed to suffer as the new technical order crashes in on them.

In summary, Winner's analysis reveals that artifacts' inherent weight manifests in various forms. Artifacts not only guide users based on designers' presuppositions, whether intentional or not. Winner (1980, p. 134) also introduces the notion that "intractable properties of certain kinds of technology are strongly, perhaps unavoidably, linked to particular institutionalized patterns of power and authority."

2.3 AI Use Within Governments

The popularity of AI has surged across various industries, driven by advancements in machine learning, data processing, and computing power. Although there are a variety of definitions of AI, the definition provided by the European Commission's High-Level Expert Group on artificial intelligence provides a clear understanding, encompassing a variety of different AI-application: "systems that display intelligent behaviour by analysing their environment and taking actions—with some degree of autonomy—to achieve specific goals" (HLEG as cited in Prins et al., 2023, p. 19). Although AI is increasingly being implemented within a variety of governmental institutions, the use of AI within government is not new. Machine vision methods, such as automatically recognizing handwriting for the United States Postal Service, were already used in the late 1990s (Mehr, 2017). Since that period, however, AI has changed, strongly expanding how AI can be utilized within governments that see merit in AI's capacity to automate routine tasks, stimulate governmental efficiency and cut costs, optimize citizen services, and facilitate governmental institutions in a variety of other ways (Engstrom et al., 2020).

Simultaneously, Big Tech companies such as IBM and Microsoft see governments and the public sector at large as a new market to promote their specifically tailored AI services. According to Big Tech, the benefits of AI are almost unlimited.

In a Google Cloud blog, head of data analytics sales at the Google public sector team Tim McLain (2024) speaks of how AI is "revolutionizing government services". Big Tech companies such as IBM, Cisco and Alphabet even invest in and contribute to new visions of government, such as the 'smart city' (i.e., cities that seek to address public issues via ICT-based solutions), in which AI tools, often provided by these same companies, are promoted as new opportunities for governments (Manville et al., 2014; Voorwinden, 2023).[3]

Big Tech and governments also claim to have an eye for the risks associated with AI and the importance of considering ethics and societal values when employing AI. Governmental institutions have a responsibility as public actors to consider the risks of AI and to employ it with respect for public values. Big Tech companies generally attempt to align their AI tools with these governmental values, at least ostensibly so, by portraying clear ethical guidelines on their website and displaying similar core values. For instance, IBM, which recently collaborated with the Spanish government to advance Spanish-language AI models, has a dedicated page displaying its ethical code and core values when it comes to AI, including values like transparency, privacy, explainability and fairness (IBM, 2024). However, recurring AI-scandals and lawsuits demonstrate that AI-related public-private partnerships also encounter challenges when it comes to achieving moral and public values. Think, for instance, about the use of facial recognition software 'Amazon Rekognition' by law enforcement agencies in the United States, software that appeared to be inaccurate and prone to racial bias (Pour, 2023).

Despite these and other examples in which safeguarding societal values when employing AI is unsuccessful, the use of AI by governments is still growing at a rapid pace. A recent report on the use of AI within Dutch public service institutions identified 50% more AI applications in 2024 compared to 2021, although a part of this growth could be attributed to increased transparency and algorithm registries (TNO, 2024). In the United States, the value of AI-related federal contracts jumped from US $355 million in 2022 to US $4.6 billion in 2023, mostly driven by the Department of Defense (Larson et al., 2024).

In practice, Colin van Noordt and Gianluca Misuraca (2022) found that AI is primarily used to improve efficiency, the delivery of public services, and enhance internal management in the European Union. In some cases, AI is also used to assist policy decision-making. In the United States, studies have shown a diversity of ways in which the federal administrative state utilizes AI, ranging from enforcing regulatory mandates and adjudicating government benefits and privileges to monitoring and analysing public health and safety risks and communicating with the public (Engstrom et al., 2020, p. 6).

[3] For more on this see *Smart cities: Private means, public ends?* by Astrid Voorwinden (2023).

2.4 AI in Government: Politicized Artifacts

Winner holds that technological artifacts can be political as they can be used to settle social or political problems, have biases towards particular social interests, and structure social and human relations. These political qualities reveal at least two ways in which the use of AI by governments should be seen as political. The first one concerns the way in which AI is designed and built. Winner's example of Moses' overpass demonstrates that, apart from *using* a technology for political or social purposes, there may also be particular social or political aims and values embedded in the *design* of the technology; aims and values that are embedded prior to the technology's immediate use. Examples in which AI systems operating on a governmental level are designed with aims and purposes that precede the direct use of AI are plentiful. Take AI chatbots, which are not only introduced to improve the quality of public services and to better communicate with citizens, but are also built to save costs by replacing civil servants. Deputy Prime Minister of the United Kingdom, Oliver Dowden, recently unveiled a 110 million pound investment plan to automate the 'dogsbody work' of civil servants (Dowden as cited in Wilcock, 2024).

Winner helps us see that values are embedded in the design of AI and that these design approaches involve choices about values that are not neutral, but political too. In line with Winner's philosophy, literature on the ethics of AI has largely stressed the importance of embedding societal and ethical values in AI-systems through methods such as (Public) Value Sensitive Design (VSD), Human-Centred Design, and Participatory Design (Degachi et al., 2024; Slingerland et al., 2024; Umbrello et al., 2021; Umbrello & van de Poel, 2021; van de Poel, 2020; Zytko et al., 2022). Others have provided theories on how AI can safeguard specific values, such as AI and Equality by Design, Transparency by Design, Privacy by Design, Human Rights by Design, and Trustworthy AI (Aizenberg & van den Hoven, 2020; Amani, 2020; Chatila et al., 2021; Dexe et al., 2020; Felzmann et al., 2020; Ravid & Hallisey, 2018). These and similar value-oriented approaches are also increasingly seen in the design of AI by governments and businesses. The Dutch government-wide vision of Generative AI, for instance, calls for a values-driven approach as a starting point to guide the development of AI (Ministerie van Algemene Zaken, 2024). Similarly, governmental institutions such as the European Commission and the United States, and Big Tech companies such as IBM, purport to develop Trustworthy AI (DOS, 2024; EC, 2019; IBM, 2024).

Despite the value of attempting to account for values in the design of AI,[4] it remains the case that values may clash and that important political choices concerning value-trade-offs have to be made. We may imagine a government that values both national security and individual privacy and is confronted with the fact that

[4] Although we fully recognize the complexity of ensuring the safeguarding of values in the ethical design of artifacts in practice, our primary objective is to demonstrate that, even when the aforementioned approaches to designing artifacts work as intended, they still undermine the value of Winner's larger contributions to understanding the politics of artifacts.

developing an AI system for security measures will likely infringe on personal privacy rights. Additionally, public and private values may clash, e.g., when the governmental use of privately-owned algorithms leads to clashes between the trade secrets of the algorithmic code and the value of governmental transparency (Carlson, 2017). In such cases, important politically laden decisions must be made about how to integrate values in the design of AI.

The focus on designing AI with values in mind, however, risks that the contestable and political nature of value-by-design approaches become ambiguous and underrecognized. As Ben Green (2021) observes, the safeguarding of values in the design and employment of algorithms often comes to be understood as being a matter of making the right technological decisions. Both Green and Winner remind us that these decisions are not a mere matter of technological reasoning, but that they are contestable and political too, and that it is important that developers of AI recognize the political positions that they are in.

There is also a second way in which Winner's concept of the politicized political quality can help us understand the politics of AI in government. In addition to the previous point, which illustrated that goals and values are purposefully being implemented within the design process of AI, Winner (1980) also shows that particular social and political goals and relations can be structured in ways that transcend thinking in terms of intended and unintended consequences (p. 125).

In these cases, the consequences and relations that accompany a technology's employment are the result of a particular constellation of social, political, scientific, technological and other factors that bias the technology in a particular direction. In such cases: "the technological deck has been stacked long in advance to favor certain social interests, and [...] some people were bound to receive a better hand than others" (ibid.). Instances of discrimination and unjust biases due to the governmental employment of AI exemplify Winner's argument. Think about recidivism-prediction tools such as COMPAS, used by prosecutors, judges, and other criminal justice actors to predict new offenses upon release, which were heavily biased against people of colour (Angwin et al., 2016; Mayson, 2018).

Notably, the use of discriminatory AI-applications is certainly not disconnected from conscious human choices about values, and so we cannot filter instances of discriminatory AI applications completely away from human intent. Nevertheless, it is certainly the case that the use of AI by governments also brings with it unexpected forms of bias and discrimination that reflect broader patterns of bias in society, as "any method of prediction will project the inequalities of the past into the future." (Mayson, 2018, p. 2218). These biases are not simply categorized as 'intended' or 'unintended'; they transcend these categories, as they are rather the result of a complex constellation of social, historical, political, technological and other factors. The unintended reflection of societal biases in such AI systems reinforces existing power structures and disparities, further disadvantaging minority communities.

2.5 AI in Government: Inherently Political Artifacts

Winner's introduction of the notion of inherently political artifacts forces us to take a closer look at AI itself. At first glance, this might sound impossible, because what qualifies as "AI itself"? The standard view in the philosophy of technology and related fields is that we should approach technology as a socio-technical system, as an artifact embedded in and in interaction with social actors. In practice, this often involves analyzing artifacts as politicized artifacts—as we have also done in the previous section—recognizing that these artifacts are shaped by actors who, to varying degrees, are positioned to influence the development and use of the application. However, what technology itself demands from its social environment in terms of institutions, governance, and other patterns of power often remains underexplored. Therefore, when looking at technology itself, we search for social and political arrangements that are necessary or at least highly compatible with a smoothly operating application. In other words, in this section, we aim to unravel what currently is seen as widely accepted "reasons of practical necessity" for AI applications to function well which overshadow political and moral reasoning.

Data

AI needs data. For AI to function, the crunching of large datasets is a sine qua non. A government that wants to reap the benefits of the so-called AI revolution must accept that it becomes a data-driven organization, investing time and resources in collecting and analyzing its citizens' data. While one can argue that information has always been a key pillar of a well-functioning government (Scott, 1998; Torpey, 2000), the current capability of collecting data from a variety of domains and combining them is unprecedented. However, this also impacts the organizational structure of the government and with it, its arrangements of authority and accountability.

Whereas a modern government is generally organized vertically, in departments, ministries, and separate organizations and institutions (e.g. police, education, health care), and with a clear hierarchical line of command, data increasingly flows horizontally, that is, between departments, ministries, and agencies. After all, it is the pooling and crunching of large data sets that fuels AI, bringing forth new predictions and insights. The promises AI applications hold urge the government to increasingly become internally connected, with data flowing from one part of the government to another.

In 2011, Prins et al. warned of the risks of not considering these new data and information flows when discussing the introduction of new applications (Prins et al., 2011, 2012). They speak of a political paradox. On the one hand, there are many political discussions and decisions resulting in the introduction of a specific application with a prime focus on its efficiency and effectiveness in solving a societal issue. On the other hand, the political dimension is utterly lacking as there is no clear legitimation of this underpinning, connected governmental system on the rise. This brings along quite some risks, such as a lack of clear stewardship of the data involved and the development of new power asymmetries between government and citizens.

Citizens become increasingly visible to the government in a way that is invisible to them, as they are not able to track how their data is being shared and used within the government (invisible visibility, see: Keymolen, 2016; Keymolen & Voorwinden, 2019). Moreover, as these information flows are also not always clear to the government either, given the lack of political debate and decision-making, taking responsibility for use and being accountable for misuse becomes more difficult.

In addition, the unquestioned acceptance of data being collected and shared as part of a smooth-functioning AI application also impacts the relationship between citizens and government. In this datafied relationship, not all citizens are equal as some become more legible than others (Dencik et al., 2019). Citizens who are dependent on government services—e.g. social benefits, healthcare, student loans—are highly present in governmental databases. While this is definitely not a new phenomenon (Scott, 1998)—after all to be able to govern one has to know who to govern—with the increasing possibilities to collect, store and analyze data, what is known about certain groups of citizens has increased substantially (Mayer-Schönberger & Cukier, 2013). Moreover, research suggests that citizens with an immigration background or lower socioeconomic status are more targeted than citizens in a higher income class (FRA, 2022). After all, data fuels AI; if there is more data on certain groups in society, it is more likely that in these domains AI applications will be developed. It should come as no surprise, therefore, that recent governmental misuse of AI applications—such as the social benefit scandal in The Netherlands and the Robodebt scheme in Australia (Berends, 2021; Royal Commission, 2023)—predominantly included more vulnerable populations. It not only harms citizens personally but most likely also has a negative impact on trust in the government, resulting in citizens no longer applying for services they are actually entitled to (Autoriteit Persoonsgegevens, 2023; Van de Walle & Bouckaert, 2003).[5] The unjust treatment of citizens is caused by both the irresponsible human use of these data-driven systems and the initial dependence on the available data that includes the high presence of citizens that are dependent on government services).

Expertise

A prevailing belief concerning AI applications is that they are complex and difficult to develop and that the government cannot do this on its own. Consequently, saying yes to AI strongly attracts public-private partnerships.

It is not new that governments outsource some of their work to external parties. Since the 1980s, with the arrival of the New Public Management ideology (Gruening, 2001; Lane, 2002), a strong preference has been nurtured to leave public tasks to contractors who presumably should be better placed than the government itself to provide services efficiently and cost-effectively. Attractive working conditions and high salaries in the tech sector further result in a lack of in-house competence to develop, build, and maintain AI applications.

[5] Distrust in government: 'Citizens have become skittish': https://nos.nl/artikel/2497916-wantrouwen-in-de-overheid-burgers-zijn-kopschuw-geworden; accessed 2-24-08-26

Big tech companies such as Microsoft and IBM dominate the AI market, becoming gatekeepers for the government to develop and use AI applications. In addition, there is a thriving market of (inter)nationally operating consultancy firms such as Deloitte, KPMG, and McKinsey that provide advice and guidance to make AI a success within government.

In theory, there should not be a problem with outsourcing work or buying applications and systems if these private providers adhere to the same public values and constitutional safeguards as the government does. However, this requires the government to clearly formulate these service conditions and, subsequently, check rigorously if they are effectively met (Freeman, 2003).

In practice, this turns out to be challenging. Public services are often "value-laden, politically contentious, and complex", leaving contractors free to "exploit interpretive gaps", implementing their own policy choices and promoting their self-interest (Freeman, 2003; Yeung, 2023: 27). Moreover, the framing of AI as a black box due to its technical complexity, combined with the intellectual property rights and trade secrecy that companies invoke, contributes to the government's loss of control over its AI applications and dependence on outside experts (Pasquale, 2015). With this dependence, governments might become weaker and even "infantilized" as some researchers claim (Mazzucato & Collington, 2023). All in all, the introduction of AI applications into the government leads to the privatization of political power.

Infrastructure

The privatization of political power is also evident in the infrastructure required for AI applications to function properly. For some time now, cloud computing has been an essential building block of most governmental digital infrastructures, and it is increasingly becoming a key component of their AI infrastructure as well. Running AI models requires significant computing power, expensive chips, and a substantial supply of electricity.

With cloud computing, essential resources such as processing capacity, databases, and storage are no longer part of the computer itself but are instead moved to the "cloud," a misleading metaphor for what is, in reality, an immense network of server farms distributed around the globe. In these packed, anonymous warehouses, long rows of buzzing machines with blinking lights process data, while ventilators run constantly to maintain a stable temperature.

AI and cloud computing converge in the possibility of choosing AI as a service (AIaaS). Rather than maintaining all resources in-house—which is expensive and as we have seen the government lacks the expertise for—one can subscribe to a cloud provider's AI services and make use of their applications, like virtual assistants and bots, their machine learning frameworks, and application programming interfaces (APIs). Seamless integration with existing systems, cost-effectiveness, and top-tier security measures are all promised as part of the package.

Gürses and Van Hoboken (2017: 9) rightfully argue that cloud computing is not merely a simple shift from the hardware capacity of users to server farms. With this move to the cloud, the control of users over their data, projects, and applications diminishes. The risk of technological lock-in also looms large. This illustrates the importance of recognizing that the inherently political nature of AI is not only located in particular functionalities and specific design choices, but also in the infrastructure that underlies the use of AI in government.

In the European Union, this move to the cloud is intertwined with a broader discussion concerning the EU and its member states' fragile digital sovereignty (see: Broeders et al., 2023; Floridi, 2020; Kaloudis, 2021). To what extent is the EU, and governments in particular, still capable of charting its course when moving to the cloud almost inevitably means entering into agreements with one of the three dominant US-based cloud providers: Amazon Web Services, Microsoft Azure, and Google Cloud Platform (Richter, 2024). Simply "taking back control" is not feasible, Timmers (2024) claims. The EU could invest in "either building up an EU cloud industry that can compete with the big foreign cloud providers (who have 70% of the market in the EU) or accepting to play a value-added role to the big cloud providers in trust services and AI with as a consequence long-term geopolitical dependency on the USA" (Timmers, 2024: 581). Although, currently cloud computing is already an accepted infrastructure within the government and is also highly compatible with the implementation and use of AI, it must be noted, that it is not the only design option available. New techniques such as edge computing, enabling AI applications to run on local devices closer to the user, could result in more decentralized cloud facilities.[6] However, whether edge computing will be "made in the EU" or also absorbed by foreign tech companies remains to be seen.

Finally, it is crucial to acknowledge that while the infrastructure of AI applications may reside in the cloud, this cloud is not virtual but material through and through. Amoore's geography of cloud forms (Cloud I) is relevant here as it focuses on the identification and spatial distributions of data centers (Amoore, 2018: 7).[7]

Currently, big tech companies such as Microsoft invest billions in setting up new data centers. Only recently the company opened its first 'cloud region' in Spain.[8] From multiple locations in the *Comunidad de Madrid*, it will provide AI and cloud services to public and private actors in Spain and Europe. As the use of AI increases, being located near a data center will become an advantage. The closer you are to where the actual computing occurs, the faster your AI application will perform.[9]

---

[6] European Commission's view on cloud computing and the role of edge computing: https://digital-strategy.ec.europa.eu/en/policies/cloud-computing; accessed 2024-08-24.

[7] It is important to note that for Amoore it is imperative that we don't get stuck in Cloud I, merely formulating critical geographical accounts of that what can be observed. She, therefore, develops Cloud II, conceptualizing how cloud computing makes activities and patterns perceptible that could never be observed directly. While we agree that Cloud II is key to grasping the depths of cloud computing, its materiality should not be overlooked as it to a certain extent pre-sorts what is rendered visible and, as we argue, also has a wide-ranging political and social impact.

[8] Press release Microsoft: https://news.microsoft.com/es-es/2024/06/11/microsoft-opens-its-first-cloud-region-in-spain-to-accelerate-the-development-of-the-ai-economy/; accessed 2024-08-16.

[9] Many new data centers for AI in the EU but not in the Netherlands: https://fd.nl/bedrijfsleven/1518931/veel-nieuwe-datacenters-voor-ai-in-europa-maar-niet-in-nederland; accessed 2024-08-16.

As these centers require substantial water and electricity supplies to operate, not all regions and countries are suitable for hosting data centers.[10] Moreover, companies face significant resistance from farmers, local residents, and environmental organizations, who fear that the arrival of data centers will lead to expropriation, the depletion of green energy, and increased noise, light, and horizon pollution.[11] In the case of the small town of Zeewolde in the Netherlands, this pushback was successful. Local opposition gained the support of the Dutch national government, leading Meta (Facebook) to permanently cancel its plans to build a hyperscale data center in the north of the country.[12]

While some view this as a success, others worry that by becoming less attractive to these big tech companies, the Netherlands risks not only missing the AI wave but also facing broader negative consequences for its local tech industry and its position as a key hub in the European digital economy.[13]

Interestingly, saying no to this necessary AI infrastructure does not immediately mean rejecting its affordances. The networked ontology of AI allows its physical infrastructure to be located far from where it is actually used. While it might become true that countries like the Netherlands, by actively resisting or simply becoming less attractive due to a shortage of green energy, could lose their favorable position in the digital economy and experience a reduction in the speed of their AI applications, they will not be deprived of AI altogether. On the contrary, they might actually avoid some of the burdens associated with AI, leaving those to other parts of the world that are not in a position to refuse this new AI industry.

Another striking example of this disconnect between use and impact is the urgent plea from Kenyan data labellers whose task is to work on safety filters for chatbots.[14] They protest the lack of support they receive on the job, which can be psychologically demanding as they are confronted with extreme content such as bestiality, suicide, torture, and murder. Despite the nature of their work, the financial compensation is extremely low (Perrigo, 2023). One could argue that "they" pay with their mental well-being for "our" safe use of AI.

All in all, the material aspects of AI, such as large data centers and the human labor involved in training AI models, can easily be overlooked when discussing its introduction in government. Saying yes to AI undeniably also means saying yes to

[10] Data center Zeewolde demands twice as much energy as the city of Amsterdam: https://www.nrc.nl/nieuws/2020/06/21/zeewolde-speelt-straks-champions-league-met-stroomverbruik-a4003548, accessed 2024-08-18.

[11] Also see the Green Web Foundation, Fieke Jansen: https://www.thegreenwebfoundation.org/news/the-politics-of-data-centers/; accessed 2024-08-18.

[12] https://datacentremagazine.com/articles/meta-permanently-cancels-plans-to-build-zeewolde-data-centre; accessed 2024-08-18.

[13] Many new data centers for AI in the EU but not in the Netherlands: https://fd.nl/bedrijfsleven/1518931/veel-nieuwe-datacenters-voor-ai-in-europa-maar-niet-in-nederland; accessed 2024-08-16.

[14] BBC report "AI and data labelling: 'I felt like my life ended'": https://www.bbc.com/news/av/world-africa-66514287; accessed 2024-08-18.

its material presence, which has a fundamental impact on our living conditions, economy, environment, and climate. This raises concerns about goodness and justice, environmental impact, and the fair distribution of the benefits and drawbacks of AI.

2.6 Discussion

As we have seen, much of the literature on designing AI with values in mind reflects Winner's concept of politicized artifacts. Similar to Winner's claim that design approaches are always accompanied with political and value-laden choices, the ethics of AI largely stresses the importance of embedding societal and ethical values in AI-systems. In addition, Winner's notion of politicized artifacts can help us understand how the governmental use of data and AI risks further entrenching existing social biases rooted in existing databases.

Moreover, we argue that Winner's philosophy about inherently political artifacts requires more attention to see what social and political structures and relations an artifact itself attracts and repels. In *Technologies as Forms of Life*, Winner (2014) warns us that overlooking how technologies restructure our physical and social world and how we live by introducing the concept 'technological somnambulism' (p. 48). He defines technological somnambulism as "willingly sleepwalking through the process of reconstituting the conditions of human existence" (ibid.), induced by both neutral (i.e., tools can be designed and used for good and bad purposes) and determinist (i.e., humans have little to no control over technology and its impact) visions of technology. Winner acknowledges that there are already important reflections on technology, including technology impact assessments and analyses of technology's side-effects. He also believes, however, that as technologies structure our world(s) in more fundamental ways, the philosophy and ethics of technology need to go beyond analyzing technology's impact and side-effects:

> The important question about technology becomes, as we "make things work," what kind of world are we making? This suggests that we pay attention not only to the making of physical instruments and processes, although that certainly remains important, but also to the production of psychological, social, and political conditions as a part of any significant technical change. Are we going to design and build circumstances that enlarge possibilities for growth in human freedom, sociability, intelligence, creativity, and self-government? Or are we headed in an altogether different direction? (Winner, 2014, p. 59)

Winner stresses that building, designing, and using technology is only one aspect of technology that requires reflection, and that more attention must be given to the ways in which these technologies fundamentally build new worlds by changing patterns of human activity and reconstructing social and political roles and relationships. Just as Winner argues that many analyses of technology lack a recognition of how technology more fundamentally restructures human worlds, the contemporary focus on how artifacts are politicized overlooks the inherently political character of technology.

Yet recognizing the inherent, political character of AI in government is of utmost importance. Employing AI within government requires a new governance structure that acknowledges the horizontal streams of data that flow through the organization. Expertise and infrastructure that is currently mainly found in the commercial, private sector, leads governmental institutions to become highly dependent on commercial parties. As AI attracts public-private partnerships, it shapes power relations by developing governmental dependence on the private sector. Similarly, analyzing the inherently political character of AI indicates the importance of recognizing the material aspects of AI, the resources it requires, and how this already structures, for instance, what locations are and are not suitable for hosting data centres. Most importantly, by recognizing the inherently political character of AI through analysing its need for infrastructure, data and expertise, we demonstrate how Winner's notion of inherently political artifacts raises new moral and political challenges for the employment of AI within government, including how to appropriately distribute its various advantages and disadvantages.

We should thus not only investigate AI's design but also explore the construction of new worlds and related social and political relations that accompany the employment of AI. An expansion of our political analyses in this regard will help us move beyond analysing how we can design specific AI applications with values in mind and will motivate us to reflect on the desirability of the particular social and political structures that the employment of AI attracts. The recent establishment of the EU AI Act is an example of a fruitful, first step in the regulation of artificial intelligence, promising to achieve a better balance between innovation, risk, and the protection of the fundamental rights of people and businesses in the EU. Whether the uptake and implementation of the AI act within governments will lead to a responsible and accountable employment of AI that recognizes AI's politicized *and* inherently political qualities in practice remains to be seen.

2.7 Conclusion

There is widespread awareness of the importance of recognizing the moral and political implications of employing AI within government. Similar to Winner's politicized artifacts, this awareness primarily manifests itself within the philosophy of technology, government, and the private sector through the popularity of value-oriented approaches and embedding societal and ethical values in AI systems. Winner's concept of inherently political artifacts, however, has not been fully integrated within contemporary discussions on the politics of AI. As a consequence, important considerations such as the increasing dependence on private parties and related political concerns about AI's use of data, infrastructure, and need for expertise, are only scarcely mentioned.

To make the responsible use of AI possible, more awareness of the inherently political aspects of AI within government and how to address these should be stimulated, for instance, by improving the ability of civil servants to recognize the inherent politics of AI. Similarly, examples of sovereign AI and cloud strategies, in which (groups of) nations (e.g., the EU) produce AI by utilizing their own infrastructure, data and expertise, reducing dependency and ensuring that such AI-systems are capable of becoming fully compatible with public values.

We also realise that developing AI systems within governments as such is easier said than done, and that it requires large investments and the unification of public/social/political values, both of which are not as readily available as we might like. Our call to increase the recognition of Winner's inherently political artifacts is therefore only a first step to understanding and dealing with the politics of AI.

References

Adadi, A., & Berrada, M. (2018). Peeking inside the black-box: A survey on explainable artificial intelligence (XAI). *IEEE Access, 6*, 52138–52160. https://doi.org/10.1109/ACCESS.2018.2870052

Aizenberg, E., & Van Den Hoven, J. (2020). Designing for human rights in AI. *Big Data & Society, 7*(2). https://doi.org/10.1177/2053951720949566

Amani, B. (2020). AI and "equality by design". *SSRN Electronic Journal*. https://doi.org/10.2139/ssrn.3734665

Amoore, L. (2018). Cloud geographies: Computing, data, sovereignty. *Progress in Human Geography, 42*(1), 4–24. https://doi.org/10.1177/0309132516673241

Angwin, J., Larson, J., Mattu, S., & Kirchner, L. (2016, May 23). Machine bias. *ProPublica*. https://www.propublica.org/article/machine-bias-risk-assessments-in-criminal-sentencing

Autoriteit Persoonsgegevens. (2023). Algorithmic risks report Netherlands. *Autoriteit Persoonsgegevens*. https://autoriteitpersoonsgegevens.nl/uploads/2023-08/Algorithmic%20Risks%20Report%20Netherlands%20-%20July%202023_0.pdf

Berends, S. (2021). *Dutch child benefit scandal: Origin and latest developments*. European Commission. https://ec.europa.eu/social/BlobServlet?docId=24723&langId=en

Boddington, P. (2017). *Towards a code of ethics for artificial intelligence*. Springer.

Broeders, D., Cristiano, F., & Kaminska, M. (2023). In search of digital sovereignty and strategic autonomy: Normative power Europe to the test of its geopolitical ambitions. *JCMS: Journal of Common Market Studies, 61*, 1261–1280. https://doi.org/10.1111/jcms.13438

Carlson, A. (2017). The need for transparency in the age of predictive sentencing algorithms. *Iowa Law Review, 103*(1). https://ilr.law.uiowa.edu/sites/ilr.law.uiowa.edu/files/2023-02/ILR-103-1-Carlson.pdf

Chatila, R., Dignum, V., Fisher, M., Giannotti, F., Morik, K., Russell, S., & Yeung, K. (2021). Trustworthy AI. In *Lecture notes in computer science* (pp. 13–39). Springer. https://doi.org/10.1007/978-3-030-69128-8_2

Degachi, C., Mehrotra, S., Yurrita, M., Niforatos, E., & Tielman, M. L. (2024). Practicing appropriate trust in human-centered AI design. In *CHI EA 24 conference, 269*. https://doi.org/10.1145/3613905.3650825

Dencik, L., Redden, J., Hintz, A., & Warne, H. (2019). The "golden view": Data-driven governance in the scoring society. *Internet Policy Review, 8*(2), 1–24. https://doi.org/10.14763/2019.2.1412

Dexe, J., Franke, U., Nöu, A. A., & Rad, A. (2020). Towards increased transparency with value sensitive design. In *Lecture notes in computer science* (pp. 3–15). Springer. https://doi.org/10.1007/978-3-030-50334-5_1

DOS. (2024, July 11). *Artificial intelligence (AI)*. United States Department of State. https://www.state.gov/artificial-intelligence/

Engstrom, D. F., Ho, D. E., Sharkey, C. M., & Cuéllar, M. (2020). Government by algorithm: Artificial intelligence in federal administrative agencies. *SSRN Electronic Journal*. https://doi.org/10.2139/ssrn.3551505
European Commission (EC). (2019). *Ethics guidelines for trustworthy AI*. European Commission. https://digital-strategy.ec.europa.eu/en/library/ethics-guidelines-trustworthy-ai
Felzmann, H., Fosch-Villaronga, E., Lutz, C., & Tamò-Larrieux, A. (2020). Towards transparency by design for artificial intelligence. *Science and Engineering Ethics, 26*(6), 3333–3361. https://doi.org/10.1007/s11948-020-00276-4
Floridi, L. (2020). The fight for digital sovereignty: What it is, and why it matters, especially for the EU. *Philosophy & Technology, 33*(3), 369–378. https://doi.org/10.1007/s13347-020-00423-6
Floridi, L., & Taddeo, M. (2016). What is data ethics? *Philosophical Transactions of the Royal Society A: Mathematical, Physical and Engineering Sciences, 374*(2083). https://doi.org/10.1098/rsta.2016.0360. https://royalsocietypublishing.org/doi/full/10.1098/rsta.2016.0360
FRA. (2022). *Bias in algorithms: Artificial intelligence and discrimination*. European Union Agency for Fundamental Rights. https://fra.europa.eu/sites/default/files/fra_uploads/fra-2022-bias-in-algorithms_en.pdf
Freeman, J. (2003). Extending public law norms through privatization. *Harvard Law Review, 116*(5), 1285–1352. https://doi.org/10.2307/1342653
Green, B. (2021). Data science as political action: Grounding data science in a politics of justice. *Journal of Social Computing, 2*(3), 249–265. https://doi.org/10.23919/JSC.2021.0029
Gruening, G. (2001). Origin and theoretical basis of new public management. *International Public Management Journal, 4*(1), 1–25. https://doi.org/10.1016/s1096-7494(01)00041-1
Gürses, S., Troncoso, C., & Diaz, C. (2011). Engineering privacy by design. *Computers, Privacy & Data Protection, 14*, 25–50.
Gurses, S., & Van Hoboken, J. (2017). Privacy after the agile turn. In J. Polonetsky, O. Tene, & E. Selinger (Eds.), *Cambridge handbook of consumer privacy* (pp. 125–142). Cambridge University Press.
Hoepman, J.-H. (2014). Privacy design strategies. In *IFIP international information security conference* (pp. 446–459). Springer. https://doi.org/10.1007/978-3-642-55415-5_34
IBM. (2024, April 5). IBM and the Government of Spain collaborate to advance national AI strategy and build the world's leading Spanish language AI models. *IBM Newsroom*. https://newsroom.ibm.com/2024-04-05-IBM-and-The-Government-of-Spain-Collaborate-to-Advance-National-AI-Strategy-and-Build-the-Worlds-Leading-Spanish-Language-AI-Models
Ihde, D. (1990). *Technology and the lifeworld: From garden to earth*. Indiana University Press.
Joerges, B. (1999a). Do politics have artefacts? *Social Studies of Science, 29*(3), 411–431. https://doi.org/10.1177/030631299029003002
Joerges, B. (1999b). Scams cannot be busted: Reply to Woolgar & Cooper. *Social Studies of Science, 29*(3), 450–457. https://doi.org/10.1177/030631299029003007
Kaloudis, M. (2021). Digital sovereignty—European Union's action plan needs a common understanding to succeed. *History Compass, 19*(1), e12698. https://doi.org/10.1111/hic3.12698
Keymolen, E. (2016). *Trust on the line: A philosophical exploration of trust in the networked era*. Wolf Legal.
Keymolen, E., & Taylor, L. (2021). Data ethics and data science: An uneasy marriage? In W. J. Liebregts, W. J. A. M. Van den Heuvel, & J. A. Van den Born (Eds.), *Data science for entrepreneurship: How entrepreneurs can leverage big data and AI for new value creation* (pp. 141–157). Springer. https://doi.org/10.1007/978-3-030-70107-8_9
Keymolen, E., & Voorwinden, A. (2019). Can we negotiate? Trust and the rule of law in the smart city paradigm. *International Review of Law, Computers & Technology, 34*(2), 1–21.
La Fors, K., Custers, B., & Keymolen, E. (2019). Reassessing values for emerging big data technologies: Integrating design-based and application-based approaches. *Ethics and Information Technology, 21*, 1–18. https://doi.org/10.1007/s10676-019-09509-x
Lane, J. (2002). *New public management*. Routledge eBooks. https://doi.org/10.4324/9780203467329

Larson, J., Dawson, G. S., Denford, J. S., & Desouza, K. C. (2024, March 26). *The evolution of artificial intelligence (AI) spending by the U.S. government*. Brookings. https://www.brookings.edu/articles/the-evolution-of-artificial-intelligence-ai-spending-by-the-u-s-government/

Manville, C., Cochrane, G., Cave, J., Millard, J., Pederson, J. K., Thaarup, R. K., Liebe, A., Wissner, M., Massink, R., & Kotterink, B. (2014). *Mapping smart cities in the EU*. Policy Department A: Economic and Scientific Policy [Report]. European Parliament. https://www.europarl.europa.eu/RegData/etudes/etudes/join/2014/507480/IPOL-ITRE_ET(2014)507480_EN.pdf

Mayer-Schönberger, V., & Cukier, K. (2013). *Big data: A revolution that will transform how we live, work, and think*. Houghton Mifflin Harcourt.

Mayson, S. G. (2018). Bias in, bias out. *The Yale Law Journal, 128*, 2218. https://heinonline.org/HOL/LandingPage?handle=hein.journals/ylr128&div=47&id=&page=

Mazzucato, M., & Collington, R. (2023). *The big con: How the consulting industry weakens our businesses, infantilizes our governments, and warps our economies*. Penguin.

McLain, T. (2024, June 25). Intelligent AI transforms government: How Google Cloud empowers public sector innovation. *Google Cloud Blog*. https://cloud.google.com/blog/topics/public-sector/powering-public-sector-innovation-with-intelligent-ai

Mehr, H. (2017). *Artificial intelligence for citizen services and government*. https://creatingfutureus.org/wp-content/uploads/2021/10/Mehr-2017-AIforGovCitizenServices.pdf

Ministerie van Algemene Zaken. (2024, January 24). Government-wide vision on generative AI of the Netherlands. *Parliamentary Document | Government.nl*. https://www.government.nl/documents/parliamentary-documents/2024/01/17/government-wide-vision-on-generative-ai-of-the-netherlands

Nyholm, S., & Smids, J. (2016). The ethics of accident-algorithms for self-driving cars: An applied trolley problem? *Ethical Theory and Moral Practice, 19*(5), 1275–1289. https://doi.org/10.1007/s10677-016-9745-2

Pasquale, F. (2015). *The Black Box Society: The secret algorithms that control money and information*. Harvard University Press.

Perrigo, B. (2023). OpenAI used Kenyan workers on less than $2 per hour to make ChatGPT less toxic. *Time*. https://time.com/6247678/openai-chatgpt-kenya-workers/

Ploug, T., & Holm, S. (2020). The four dimensions of contestable AI diagnostics: A patient-centric approach to explainable AI. *Artificial Intelligence in Medicine, 107*, 101901. https://doi.org/10.1016/j.artmed.2020.101901

Pour, S. (2023). Police use of facial recognition technology and racial bias: An assessment of criticisms of its current use. *American Journal of Artificial Intelligence*. https://doi.org/10.11648/j.ajai.20230701.13

Prins, C., Broeders, D., Griffioen, H., Keizer, A.-G., & Keymolen, E. (2011). *iGovernment*. Amsterdam University Press.

Prins, C., Sheikh, H., & Schrijvers, E. (2023). *Mission AI: The new system technology*. Springer. https://doi.org/10.1007/978-3-031-21448-6

Prins, J. E. J., Broeders, D., & Griffioen, H. M. (2012). iGovernment: A new perspective on the future of government digitisation. *Computer Law & Security Review, 28*(3), 273–282. https://doi.org/10.1016/j.clsr.2012.03.005

Richter, F. (2024, May 2). Amazon maintains cloud lead as Microsoft edges closer. *Statista Daily Data*. https://www.statista.com/chart/18819/worldwide-market-share-of-leading-cloud-infrastructure-service-providers/

Royal Commission. (2023). *Report of the Royal Commission into the Robodebt Scheme*. https://robodebt.royalcommission.gov.au/system/files/2023-09/rrc-accessible-full-report.PDF

Scott, J. C. (1998). *Seeing like a state: How certain schemes to improve the human condition have failed*. Yale University Press.

Slingerland, E., Ongena, G., & Van Steenbergen, M. (2024). Developing public values-based AI systems using value sensitive design. In *37th Bled eConference*. https://doi.org/10.18690/um.fov.4.2024.50

Timmers, P. (2024). Sovereignty in the digital age. In H. Werthner, C. Ghezzi, J. Kramer, J. Nida-Rümelin, B. Nuseibeh, E. Prem, & A. Stanger (Eds.), *Introduction to digital humanism: A textbook* (pp. 137–149). Springer. https://doi.org/10.1007/978-3-030-81455-7_10

TNO. (2024, June 24). Steeds meer kunstmatige intelligentie ingezet door overheid. *TNO.nl*. https://www.tno.nl/nl/newsroom/2024/06/kunstmatige-intelligentie-inzet-overheid/

Torpey, J. (2000). *The invention of the passport: Surveillance, citizenship, and the state*. Cambridge University Press.

Umbrello, S., Capasso, M., Balistreri, M., Pirni, A., & Merenda, F. (2021). Value sensitive design to achieve the UN SDGs with AI: A case of elderly care robots. *Minds and Machines, 31*(3), 395–419. https://doi.org/10.1007/s11023-021-09561-y

Umbrello, S., & Van de Poel, I. (2021). Mapping value sensitive design onto AI for social good principles. *AI and Ethics, 1*(3), 283–296. https://doi.org/10.1007/s43681-021-00033-7

Vallor, S. (2016). *Technology and the virtues: A philosophical guide to a future worth wanting*. Oxford University Press.

Van De Poel, I. (2020). Embedding values in artificial intelligence (AI) systems. *Minds and Machines, 30*(3), 385–409. https://doi.org/10.1007/s11023-020-09537-4

Van der Hof, S., & Lievens, E. (2018). The importance of privacy by design and data protection impact assessments in strengthening protection of children's personal data under the GDPR. *Communications Law, 23*(2), 93–103.

Van de Walle, S., & Bouckaert, G. (2003). Public service performance and trust in government: The problem of causality. *International Journal of Public Administration, 29*(8–9), 891–913. https://doi.org/10.1081/PAD-200066169

Van Noordt, C., & Misuraca, G. (2022). Artificial intelligence for the public sector: Results of landscaping the use of AI in government across the European Union. *Government Information Quarterly, 39*(3), 101714. https://doi.org/10.1016/j.giq.2022.101714

Verbeek, P.-P. (2011). *Moralizing technology: Understanding and designing the morality of things*. University of Chicago Press.

Voorwinden, A. (2023). Smart cities: Private means, public ends? *Technological Forecasting and Social Change, 182*, 121899. https://doi.org/10.1016/j.techfore.2023.121899

Winner, L. (1980). Do artifacts have politics? *Daedalus, 109*(1), 121–136. https://www.jstor.org/stable/20024652

Winner, L. (2001). Where technological determinism went. In S. Cutcliffe & C. Mitcham (Eds.), *Visions of STS. Counterpoints in science, technology, and society studies* (pp. 11–18). State University of New York Press.

Winner, L. (2021). The democratic shaping of technology: Its rise, fall, and possible rebirth. *Engaging Science, Technology, and Society, 7*, 95–103. https://doi.org/10.17351/ests2021.999

Woolgar, S., & Cooper, G. (1999). Do artefacts have ambivalence? Moses' bridges, Winner's bridges and other urban legends in S&TS. *Social Studies of Science, 29*(3), 433–449.

Yeung, K. (2023). The new public analytics as an emerging paradigm in public sector administration. *Tilburg Law Review, 27*(2), 1–32.

Zytko, D., Wisniewski, P. J., Guha, S., Baumer, E. P. S., & Lee, M. K. (2022). Participatory design of AI systems: Opportunities and challenges across diverse users, relationships, and application domains. In *CHI conference on human factors in computing systems extended abstracts*. https://doi.org/10.1145/3491101.3516506

Chapter 3
Artifacts' Agency | Designers' Agency

Meredith Sattler

Abstract This chapter interrogates Langdon Winner's now infamous piece "Do Artifacts Have Politics" through the lens of Design. As both designers and users of technologies and infrastructures, there is much to be gained from disciplinary translations and deeper considerations of Winner's ideas: refocusing on contextual and situated aspects surrounding a designed object or system, foregrounding issues related to time and obduracy, encouraging ethical Design considerations, and increasing the scope of designers' and users' agency, among others. Yet, Winner's work, largely discussed from a Sociological and more collective standpoint, requires translation into Design Praxis. To accomplish this, the chapter explicates some of the utility, and inherent tensions, in the scales and temporal positions of Sociological epistemologies and ontologies of collectives, for designers and their users, at work. In so doing, it further unpacks how designers may utilize Winner's relational frameworks of artifacts and politics to more consciously understand, pluralize, leverage, and predict potential implications of their impacts in the world. Divided into two parts, the chapter first contextualizes Winner's argument in relation to Design, while the second proposes three design tools/approaches that bridge Science and Technology Studies and Design praxis, which may assist designers to better forecast the politics of their products.

Keywords Artifacts · Politics · Design · Design Process · Agency

---

M. Sattler (✉)
Architecture Department, California Polytechnic State University,
San Luis Obispo, CA, USA

© The Author(s), under exclusive license to Springer Nature Switzerland AG 2026

A. Fox, D. Cressman (eds.), *Arrangements of Power: Tracing Langdon Winner's Legacy Within and Beyond the Philosophy of Technology*, Philosophy of Engineering and Technology 50, https://doi.org/10.1007/978-3-032-17371-3_3

3.1 Introduction

Langdon Winner's now infamous piece "Do Artifacts Have Politics,"[1] and its subsequent debates,[2] have not only endured the test of time, but have increasingly found their way into core extra-disciplinary conversations, including those of Design. Winner's proposition, that the social may be reconfigured through the intentional design of artifacts, is simultaneously empowering to designers of these artifacts and daunting. Designers, typically poorly educated in the political, let alone the social, find themselves in precarious positions when the products of their work (often to their surprise) are framed as technologies that are "…'solution[s]' to political order in the sense that [they] sustain particular structures of established power…"[3]

As Design educators, there is much to be gained from introducing Winner's ideas in the Design Studio environment: bringing attention to the contextual and situated aspects surrounding a designed object or system, foregrounding issues related to time and obduracy, encouraging a heightened awareness/ urgency of ethical considerations in design, and increasing the scope of the agency of designers' work, among others. Yet, in this context there are also necessary translations from the implications of Winner's work, largely discussed from a Sociological and more collective standpoint, to the point of view of Design Praxes, or designers' decision-making processes, often in direct relation to their client's prerogatives and their own disciplinary aims. This chapter explicates some of the utility, and inherent tensions, in this 'translation' from scales and temporal positions of Sociological epistemologies and ontologies of collectives to the position of the individual designer and/or their team, at work. In doing so, it attempts to further unpack how designers may utilize Winner's relational frameworks of artifacts and politics to more consciously understand, pluralize, leverage, and predict potential implications of their impacts in the world.

To accomplish this, this chapter is divided into two parts: the first contextualizes "Do Artifacts Have Politics," in relation to Design and the designer, and the second proposes three design tools/approaches that bridge Science and Technology Studies [STS] and Design Praxis, which Winner doesn't directly address yet may assist designers to better forecast the politics of their products. The chapter begins Part I by briefly tracing the history of Winner's engagement with the discipline of Design and his text "Do Artifacts Have Politics" before situating Winner's work in relation to contemporary Design contexts, theories, and praxis. It then reads his text through the lens of Design, summarizing his position and interests, and drawing attention to

---

[1] Winner, L. (1980). "Do Artifacts Have Politics?" *Daedalus, 109*(1), 121–136.

[2] Joerges, B. (1999). "Do Politics Have Artifacts?" *Social Studies of Science,* 29(3), 411–431. Woolgar, S. & Cooper, G. (1999). "Do Artifacts Have Ambivalence?" *Social Studies of Science,* 29(3), 433–449. Joerges, B. (1999). "Scams Cannot be Busted" *Social Studies of Science,* 29(3), 450–57. Among others.

[3] Jasanoff, S., (Ed.). (2004). *States of Knowledge: The Co-Production of Science and the Social Order*. Routledge, 30.

some of the tensions inherent within the Design Processes. Next, it unpacks relationships between temporality and agency inherent in the designer's process which can be difficult to reconcile with a more classical Socio-Political view of artifacts, and vice versa. This leads to consideration of arguments made within the 1999 debate that appeared within *Social Studies of Science*, between Bernward Joerges, Steve Woolgar, and Jeff Cooper, on Winner's depiction of Robert Moses's low-slung Long Island Parkway(s) bridges as a way to reexamine some of the difficulties translating the *language* of Sociology into the *materiality* of Design. Finally, Part I concludes with an analysis, inspired by the 1999 debate, that traces how Winner's assignment of political agency to artifacts forces reconsideration of both the roles of 'designers,' and of those implicated in the production and use of their artifacts and systems, and how this reconsideration simultaneously implicates categorizations we typically assign to artifacts themselves. In Part II, the chapter builds upon this reconsideration by offering three Science and Technology Studies related analytical frameworks that are particularly accessible for 'designers' to leverage in analysis of the political implications of their products within their Design Processes: Scripting or In-Scription,[4] Affordances,[5] and Systems Oriented Design.[6]

3.2 Part I: "Do Artifacts Have Politics" and Design

3.2.1 Langdon Winner, Design, and "Do Artifacts Have Politics"

> If our moral and political language for evaluating technology includes only categories having to do with tools and uses, if it does not include attention to the meaning of the designs and arrangements of our artifacts, then we will be blinded to much that is intellectually and practically crucial. (Winner, 1986, 25)

Langdon Winner's seminal 1980 essay, "Do Artifacts Have Politics," remains one of the most cited works in STS, in addition to receiving considerable recognition across multiple disciples. Winner's text explores the intersection of technologies, politics, and society, with a primary focus on the ways in which technological artifacts influence power structures, societal organization, and human behavior. While his primary field is STS, Winner's work has profound implications for the field of Design. It introduces designers to the notion that their artifacts may contain intrinsic

[4] Akrich, M. & Latour, B. (1992). "A Summary of a Convenient Vocabulary for the Semiotics of Human and Nonhuman Assemblies." In W. Bijker & J. Law (Eds.) *Shaping Technology, Building Society: Studies in Sociotechnical Change*. MIT Press, 259–264.

[5] Norman, D. (2013). *The Design of Everyday Things, Revised and Expanded Edition*. Basic Books. and Gibson, J. (1979). *The Ecological Approach to Visual Perception*. Psychology Press (2015 "classic" edition).

[6] Sevaldson, B. (2022). *Designing Complexity: The Methodology and Practice of Systems Oriented Design*. Common Ground.

politics embedded within them, and as a result, challenges designers to consider the political consequences of their creations, both intended and unintended. Ultimately, Winner's work has been used in many pedagogical and professional contexts to compel designers to critically reflect on the social implications of their work and to rethink their role in shaping collective futures.

Winner's text can be understood as not only laying groundwork for the Social Construction of Technology,[7] forecasting the emergence of New Materialisms (particularly Jane Bennett's Vibrant Matter),[8] and even influencing the development of early Posthuman frameworks such as Actor Network Theory [ANT][9] and Material Semiotics,[10] but is also part of a larger tradition in Design Theory that calls for a more reflexive approach to praxis. The rise of Critical Design in the 1990s and early 2000s, perhaps launched by the introduction of the concept of *Wicked Problems*[11] in 1973 by Rittel and Webber at Berkeley,[12] emphasizes the interrogation of societal norms through Design. Critical Design certainly draws on Winner's exploration of intersections of technology and politics such that designers are encouraged to not just see their work as part of a larger system and to solve problems within it, but to interrogate the political dimensions of those problems, acknowledging the potential for design to reinforce or subvert power dynamics while simultaneously addressing the uncertainty that inherently accompanies the release of their products into larger political, economic, and social milieus. *Wicked Problems* takes these considerations to their extreme, by disputing key assumptions within the Design disciplines, namely the homogeneity of society/user groups in tandem with a damning critique of *problem definition*, both of which render more traditional 'positivist' Design Solutions impossible. Instead, Rittel and Webber argue that Design should embrace less deterministic approaches that complicate Winner's position which relies on a causal Design trajectory where a political figure's intent (Robert Moses) is systematically designed, materialized, and manifest through the use of an infrastructure (parkway bridges).

Though Winner's work consistently demonstrates an interest in not just the politics of technologies, but also their genesis within Design (he taught within the Rensselaer Polytechnic Institute's joint programs between STS and Design for

[7] Bijker, W., Hughes, T. & Pinch, T. (Eds.). (1987). *The Social Construction of Technological Systems*. The MIT Press.

[8] Bennett, J. (2010). *Vibrant Matter: A Political Ecology of Things*. Duke University Press.

[9] Latour, B. (2005). *Reassembling the Social: An Introduction to Actor-Network-Theory*. Oxford University Press.

[10] Law, J. (2023). "Material Semiotics." In Atkinson, P., Delamont, S., Cernat, A., Sakshaug, J. W. & Williams, R. A. (Eds.), *SAGE Research Methods Foundations*. Sage. and Harley, A. (2019). "Haraway's Material-Semiotic Knot: A Learning-Teaching Response for Creative-Critical Times." *a|b: Auto|Biography Studies, 34*(3): 539–553.

[11] Rittel, H. & Webber, M. (1973). "Dilemmas in a General Theory of Planning." *Policy Sciences, 4*(2): 155–169.

[12] Perhaps not coincidentally, Rittel and Webber, both Berkeley Professors, published their text that introduced the concept of Wicked Problems in the same year that Winner received his PhD, also from Berkeley.

decades), his background, and positionality, are firmly rooted in Political Theory, having completed his B.A., M.A., and PhD in Political Science at Berkeley from the mid-1960s to 1970s. As alluded to above, at the time Berkeley's zeitgeist was immensely influential for Winner, instilling his keen interests in social and environmental justice and activist-leaning approach. So, it is not surprising that Winner's interests led him to orbit through the Design of technologies, despite his lack of Design training or Praxis.

3.2.2 The Governmentality of Things *in "Do Artifacts Have Politics"*

In Winner's central argument in "Do Artifacts Have Politics," technological artifacts cannot be 'neutral.' Instead, they are always imbued with governing (read as 'political') properties that structure human relations and behaviors in profound ways. I argue that the distinction between 'agency' and 'governing' in this context is critical, as Winner's politics of artifacts require more power than just 'agency' alone might suggest, particularly when seated within larger technological systems. This use of 'governing' then allows an alignment with Foucault's work on 'government,' which Lemke's 2021 reassessment of his 1977–1978 "Security, Territory, Population" lectures directly supports. Lemke argues that Foucault's governmentality does not just encompass the political and social concerns of ethics and subjectivation, as it is typically interpreted, but instead includes strong entanglements between the social and the material: "a government of things makes it possible to arrive at a relational and strategic understanding of agency and ontology that productively engages with some of the issues the new materialisms raise [including entanglements of humans and nonhumans]." (Lemke, 2021, 10–11).

Winner is particularly interested in forms of authoritarian or democratic power that manifest through objects—leveraged depending on how they are designed, implemented, and used—complement Foucault's assessments. Winner structures his analysis along two intersecting lines: the *Politics of Design* and the *Politics of Technological Systems*. For Winner, the Politics of Design are the shaping of social relations through the materiality and structure of artifacts, whether intentionally or unintentionally. His Politics of Technological Systems situates these artifacts within larger technological infrastructures, often imbued with centralized or rigid hierarchical power structures, that may exert broader governing dynamics. In both cases, designers may play key roles in shaping how designed interventions exert agencies—though as their designs increase in scale and complexity, the ability for designers, and by implication their clients, to control their intentional outcomes diminish.

Here, several concepts from The Social Construction of Technology [SCoT] are helpful in unpacking mechanisms that contribute to the realities of designer's roles and agencies in their process.

Wiebe Bijker's *Technological Frame* is a concept

> ...intended to apply to the interaction of various actors... The technological frame of [a] social group structures attribution of meaning by providing, as it were, a grammar for it. This grammar is used in the interactions of members of that social group, thus resulting in a shared meaning attribution... A technological frame is built up when interaction 'around' an artifact starts and continues...On the one hand, a technological frame can be used to explain how the social environment structures an artifact's design...On the other hand, a technological frame indicates how existing technology structures the social environment. (Bijker, 1987, 168–169)

This author argues that there may be a third function of Technological Frames, where designers' intentions may be imbued into designed artifacts (and to some extent even their surrounding contexts), either unconsciously, or more deliberately. Essentially, designers may either leverage an existing social groups' 'grammar,' or attempt to create a new 'grammar' communicated by their artifact, via mechanisms that the three design approaches introduced in the second part of this chapter elucidate. However, they still often cannot completely control the reading of this 'grammar,' or alternate uses of their artifact.

Additionally, the STS concept of *Interpretive Flexibility* demonstrates how shared frames of meaning may shift through use over time. As Pinch and Bijker discuss in their 1987 text "The Social Construction of Facts and Artifacts: Or How the Sociology of Science and the Sociology of Technology Might Benefit Each Other," Interpretive Flexibility means that each technological artifact may have different meanings and interpretations for different user groups. This means "...that there is flexibility in how people think of or interpret artifacts..."[13] These alternative interpretations may shift through time, as Technological Frames evolve, quite independently of the designer's initial framing, and in so doing may generate different problems to be solved, resisting *Closure*, or the *Stabilization* of an artifact's use and/or meaning. These concepts challenge Winner's generally deterministic perspective on how the design of artifacts may contain targeted political agendas that determine use—specifically, his assessment of Robert Moses's potential intent, as manifest through the design of low-slung bridges as gatekeeping mechanisms to keep bus-riding people of lower socio-economic classes out of Jones Beach, New York.

3.2.3 *The Designer's Process: Temporality and Agency*

Temporal trajectories inherent in the design process further complicate Winner's assessment and how it troubles the designer's agency in several ways: the design of particular artifacts and technological systems occurs within particular

[13] Bijker, W. & Pinch, T. (1987). "The Social Construction of Facts and Artifacts: Or How the Sociology of Science and the Sociology of Technology Might Benefit Each Other." In W.E. Bijker, T.P. Hughes and T. Pinch (Eds.), *The Social Construction of Technological Systems.* The MIT Press, 34.

circumstances at particular times, and hence, are conditioned by these prevalent influences. Designers, and/or their clients, determine the needs that designed objects and/or systems should address, as understood at that time, to the best of their ability. They then proceed to develop Design Criteria that ultimately guide the Design Process toward primarily addressing those needs. However, as Bijker and Pinch point out, "there is flexibility in how artifacts are designed. There is not just one possible way or one best way of designing an artifact."[14] So, even within Winner's Politics of Technological Systems framework, where top-down control is intentionally and rigorously designed into the system, and dominates its use, or as Jasanoff describes, "technology in these terms is a 'solution' to political order in the sense that it sustains particular structures of established power"[15]—the unexpected still emerges.

This can happen in many ways, as the STS Literature on Ignorance Studies,[16] Surprise,[17] and the production of Undone Knowledge[18] traces. Or, very commonly in Design Praxis, due to the largely standardized phases of the Design Processes, and their temporal frames. For example, many design workflows do not emphasize consideration of use beyond a certain time-horizon, or changing uses through time, (which relates to the complications discussed in shifting meanings and uses throughout this chapter). For example, most U.S. based Architectural pedagogies don't emphasize serious consideration of buildings much past their initial occupation. This results in a problematic bias throughout the Design Process which renders these very obdurate built-artifacts essentially timeless, unaffected by aging or weathering considerations, often to the future frustration of their client/owners. Ironically, Steward Brand's work on the life-spans of different sub-building systems in his 1994 *How Buildings Learn*, has been enlightening to Architects who encounter it, and who otherwise rarely consider how these often-enmeshed building components will need to be replaced at different times, with great difficulty, but could be initially designed to accommodate this almost universal need instead.[19]

Further, as both Technological Frames and Interpretive Flexibility demonstrate, artifacts may be significantly more physically durable through time than the uses, memories, and meanings associated with them. Likewise, as Lemke emphasizes, "Politics [are] not a given, stable and self-evident entity; rather, [their] contingent boundaries and material conditions come to the fore."[20] So it may very much be the

[14] *ibid*: 34.

[15] Jasanoff, 30–31.

[16] Gross, M. & McGoey, L. (2015). *Routledge International Handbook of Ignorance Studies*. Routledge.

[17] Gross, M. (2010). *Ignorance and Surprise: Science, Society, and Ecological Design*. The MIT Press.

[18] Hess, D. (2016). *Undone Science: Social Movements, Mobilized Publics, and Industrial Transitions*. The MIT Press.

[19] Brand, S. (1994). *How Buildings Learn: What Happens After They're Built*. Penguin Books, 35.

[20] Lemke, T. (2015). "New Materialisms: Foucault and the 'Government of Things.'" *Theory, Culture & Society, 32*(4): 16.

case that the design of a more durable artifact is accomplished within one Technological Frame, but that its obduracy spans several Technological Frame iterations, far superseding the designer's and/or politician client's intent. The fact that many designed artifacts and systems physically long outlive their initially intended politics, and even uses, complicates the agency of those involved in the relatively deterministic temporal trajectory of the design and construction process. Ultimately, designers may grapple with the very challenging, if not impossible, task of predicting, let alone designing, for futures that are highly indeterminate.

Finally, the temporal inversions inherent in Design Praxis versus the study/analysis of design products affect the understanding of designed artifacts performance for different groups. The difference between a retroactive (Sociological analysis or thinking from artifact to design intent) and a proactive (Design Praxis or thinking from design intent to artifact) approach to analyzing Winner's argument, which is deeply seated in Sociological traditions, requires that the designed artifact exists and has interfaced with the larger cultural milieu, to be observed and analyzed. This process contrasts those of designers who must work in a literally constructivist-projective mode, where their analysis may be informed by observation of past case studies or 'precedents' of how comparable designs have performed, but by necessity must largely be based on forecasted future scenarios. As such, Design Praxis is more contingent and reliant on the experiences of their practitioners making educated guesses and testing the possible outcomes through their 'practice,' which is inherently a more uncertain endeavor. Essentially, it can be more difficult for designers to come to conclusions about the performance of an object; whereas, it can be more difficult for Sociologists to come to conclusions about the generation of an object. This sheds new light on the debate discussed in the next section.

3.2.4 Language *and* Materiality *in the 1999* Social Studies of Science *Exchange*

In 1999, almost 20 years after "Do Artifacts Have Politics" was first published, Winner's piece spawned a debate in the *Social Studies of Science* journal, between Bernward Joerges, Steve Woolgar, and Geoff Cooper. In the end, ironically, the thrust of the argument revolved around the verbal communication of meaning, not materiality, and loosely socio-political agendas. Given that it was near the end of the Science Wars, when constructivism, postmodernism, and objectivity were largely debated, perhaps it's not entirely surprising.

Joerges initiated the exchange by arguing that Winner's work misinterpreted the functionality of the low-slung bridges, as evidenced by NYC Bus routes and schedules from Manhattan to Jones Beach, thereby invalidating his argument based on the physical agency of the bridges. In its place, Joerges came to the conclusion that Winner's analysis was more about making a discursive argument than one about the politics of artifacts, one that revolved around "…how to do things with words,' exactly not about 'how to do things with things.'"[21] He goes on to argue that "Winner's Moses example is so winning because in itself it is a particularly well constructed artifact, capable of serving a great number of rhetorical purposes."[22] Woolgar and Cooper follow-up, with the conclusion that:

> ...Winner's bridges are also artefacts constructed with the intention of not letting certain arguments past. They are deliberately designed to prevent the passage of interpretivist arguments. (Woolgar & Cooper, 1999, 444)

Finally, Joerges concludes the debate by circling back from the discursive, toward the material, through forecasting arguments that would later gain traction in New Materialists conversations:

> In the future, practical and interesting answers to Winner's ultimate moral concern—what kind of nature should you make yourself?—may be informed more and more by expertise kindled from reductionist evolutionary sciences claiming society as their native land. This is a good thing, to the extent that 'nature,' and material artifacts taken from nature, cannot any longer be represented as the Other of society. (Joerges, 1999, 455)

While Joerges reasserts that material objects are social actors, much in the way that Latour's ANT attempts to dissolve boundaries between humans and non-humans,[23] the debate ultimately reorganizes the matter, meaning, and politics of Winner's view of Moses's bridges.

3.2.5 Reassigning Agency | Renegotiating Categories

The 1999 debate foregrounds how language's inherent categories may contain as much agency as material artifacts themselves, or potentially even more, circling back to the earlier Technological Frames discussion of different 'grammars' used by different social groups.

Through further analysis of the implications of Winner's Politics of Design, larger questions emerge around classification of the roles of 'designers' and 'users,' and the ways these roles construct meanings, identities, and practices. If in Winner's framing, politicians can author materializations that shape social relations through 'client-directives' to designers, he opens the door for policymakers themselves to be understood as designers. By further 'tracing the agency' through enlarged scopes of Design/implementation practices, it becomes clear that those in charge of implementation, and even compliance, could be understood as performing design roles, and might therefore be identified as designers of the social realm. For example,

---

[21] Joerges, 415.

[22] *ibid*, 420.

[23] Latour, B. (2005). *Reassembling the Social: An Introduction to Actor-Network-Theory*. Oxford University Press.

within more traditional architectural or product design scope, this might look like General Contractors or Construction Workers interpreting how a construction/shop drawing specification is built on-site, or Manufacturers determining how a factory production run is tooled and staffed.[24] Further, municipality Building Inspectors signing-off on code compliance or issuing occupancy permits, or Quality Control Operators performing final product inspections might also be considered designers due to their direct influence on how products are materialized, made available for distribution, and used. Delving deeper into the cultural by including political, economic, and informational realms, this re-scoping might contain many more actors who influence infrastructural and/or systemic social structures through regulation, or via the governing forces latent in the structures of information and large infrastructural systems architectures (more on this below).

Equally, Winner's Politics of Design encourages additional inquiry into how we understand categories of 'users' through both the use of SCoT's analytical framework of Interpretive Flexibility, and via reexamination of the categories that structure the analysis of Micro and Macro-Economic market functionality. As previously stated, Interpretive Flexibility reveals how users may reinterpret and use a designed artifact in very different ways than intended by design, which this author argues expands and redesignates the 'user' into a 'user-designer.' For example, the road sign documented below, when not just read as a warning to drivers, clearly depicts the design agency of a Bus Driver who might intentionally choose to read the sign as a set of instructions for reconfiguring both their vehicle and the bridge, by driving in the right lane (Fig. 3.1).

The above 'cyborg assemblage,'[25] and others that more intimately entangle humans and non-humans, such as those examined by Lucy Suchman in her book *Human-Machine Reconfigurations*,[26] redistribute agency between human 'users,' and other partially pre-programmed or structured inanimate products, that only function through hybrid-use (see section on Affordances in Part II). These artifact-assemblages additionally confuse functional designations of 'client,' 'designer,' 'producer,' 'product,' 'customer,' and 'consumer.' Here designers' 'clients' and 'consumers,' whom in more traditional Design Process trajectories clearly occupy only the beginning and end of the process respectively, may become 'designer-producers' within multiple points along that trajectory. Similarly, even 'customers' may become 'designers' in the way that Marcel Duchamp redefined the category of the Artist as one who 'makes' art through acts of selection, or 'shopping,' using his

---

[24] Noble, D. (1984). *Forces of Production: A Social History of Industrial Automation*. Alfred A. Knopf.

[25] Haraway, D. (1985). "Manifesto for Cyborgs: Science, Technology, and Socialist Feminism in the 1980's." *Socialist Review* 80: 65–108.

[26] Suchman, L. (2007). *Human-Machine Reconfigurations: Plans and Situated Actions, 2nd Edition*. Cambridge University Press.

Fig. 3.1 Road signage on George Washington Memorial Parkway, Fort Hunt, VA. Photograph by the author

"Readymades" production process.[27] For designers, all of the above examples might catalyze serious reconsideration of their role(s) within, and approach(es) to Design.

In expanding this analysis into the implications of Winner's Politics of Technological Systems, the specific mechanisms of governmentality within Winner's Technological Systems, and those of Thomas Hughes's related Large Technological Systems [LTS],[28] must be further unpacked. Winner defines his

[27] Readymades is "... a term coined by Marcel Duchamp in 1916 to describe prefabricated, often mass-produced objects isolated from their intended use and elevated to the status of art by the artist choosing and designating them as such. The term 'assisted Readymade' refers to works of this type whose components have been combined or modified by the artist." MOMA. 2025. "Readymade." Accessed March 18, 2025. https://www.moma.org/collection/terms/readymade

[28] Winner's *Politics of Technological Systems* are not dissimilar to Thomas Hughes analytical framework of Large Technological Systems [LTS] first published 3 years after Winner's in:

Technological Systems as politicized assemblages of artifacts structured into centralized, tightly controlled infrastructural relationships organized according to rigid hierarchies. He provides numerous examples of how these systems intentionally exert governing influences that reinforce top-down power dynamics. Here it is important to interrogate Winner's assumption that these systems are always intentionally imbued with these top-down dynamics. As seen in early histories of the World Wide Web, LTS's may or may not be structured by centralized hierarchical architectures. As such, the DARPA version of the early internet was largely conceived as a tightly controlled military-industrial communications technology, whereas "The Well" version was a much more 'bottom-up' assemblage that aspired to operate completely free of hierarchy, in the Hippie spirit with which it was created. Fred Turner's work argues that these communication infrastructures reflect the social ideals of their creator-users,[29] thereby demonstrating that nearly identical LTS technologies may be structured and operate in vastly different ways across multiple sectors of influence.

Regardless of whether a specific power dynamic orients a System's Architecture, it is clear that the structure of that system's organization produces governing forces that control users' ability to access and engage with the system's components and desired services. These affect not only their intended users, but simultaneously unintended users, designers' and their clients' abilities to reconfigure the system, and maintainers' abilities to operate and repair the system, among others. Each of these, in turn, not only configure workflows and potential products, but simultaneously their related social dynamics.

As seen above, the analytical work of unpacking Winner's narrative from the materialist side is markedly different from that from the social side, which is fascinating and in need of more attention. In the next section, the author attempts to further bridge between the Sociological and Design disciplines from the material perspective, by introducing three STS related analytical tools that have been constructed with material considerations in the foreground. These may assist designers in integrating Winner's concerns more systematically into their Design Thinking and Praxis.

Hughes, T. (1983). *Networks of Power: Electrification of Western Society, 1880–1930*. John Hopkins University Press.

[29] Turner, F. (2006). *From Counterculture to Cyberculture: Stewart Brand, The Whole Earth Network, and the Rise of Digital Utopianism*. University of Chicago Press.

3.3 Part II: Design Tools/Approaches That Facilitate Unpacking Artifact's Politics

Winner's intent that the analysis and study of intersections between technologies and social theory should enable new and more informed approaches to Design is evident, though perhaps backgrounded, in many of his texts. In order to more directly link Winner's reading of technology to the Design disciplines, a linguistic translation is required—another type of *doing of things with words,* that may facilitate a *doing of things with things*.[30] By leveraging a definition of the term 'technology' that *Oxford English Dictionary* traces to a 1858 origin: "a particular practical or industrial art; a branch of the mechanical arts or applied sciences; a technological discipline,"[31] this author argues that almost all products originating from generative design or 'technological art' processes produce technologies, in the broadest sense. For example, houses can be understood as technologies, as well as medical diagnostic apparatuses, as well as Systems Architectures and even the Process Diagrams that generate and represent them. Hence, we can now better read how Winner more directly addresses the impact of his social work on Design Processes, through the way he frames his courses taught in the Rensselaer Polytechnic Institute's joint programs between STS and Design.

For example, in a 2004 seminar offering on "Technology and Social Theory" Winner's three course objectives can be summarized as: (1) conduct close readings of works in history, literature, philosophy, and social science that seek to illuminate our understanding of technology's place in human experience; (2) develop a grasp of key themes and issues in a manner compatible with each student's view of social theory and STS (which accommodates students of Design's views); and (3) apply this knowledge to contemporary social and political issues that involve technological choice (which may include choices about both design and use).[32] He goes on to suggest possible student projects which should examine "...contemporary social or political issue[s] related to existing technolog[ies] or pending technological choice[s]...[such as] manufacturing technology and the definition of work; global production and its consequences;...the philosophy and practice of sustainable development...[and] theoretical reflections on a contemporary sociotechnical landscape [such as]: computer networks, megamalls...high-performance workplaces, gated 'communities,' material cultures of the homeless, etc."[33] These approaches, and more, require STS and Sociological frameworks with which to conduct analysis of designed technologies, circumstances, or environments. These may be further enhanced by the following analytical frameworks that bridge through STS, and

[30] Joerges, 415.

[31] *Oxford English Dictionary*. 2024. "technology (n.), sense 5," September, https://doi.org/10.1093/OED/4236248164.

[32] Winner, L. (2004). "Technology and Social Theory." Syllabus, Fall, Rensselaer Polytechnic Institute, 1.

[33] *ibid*, 2.

allied disciplines, into Design Processes, which are particularly useful for designers and those engaged in developing and testing their output via the constructivist-projective mode of Design Production.

3.3.1 Scripting or In-Scripting

The operations of *Scripting*, *Inscription*, *De-scription*, and *Transcription* are at the core of STS analysis of the Semiotics of Human and Non-Human Assemblies and are rendered especially evident within the context of ANT and its related approaches.[34] Originally developed as STS analytical tools, they may simultaneously be utilized as projective tools by designers to predict/test iterative outcomes of their Design Process. Scripting allows designers to design a *program of actions* as a "... text of what the various actors in the setting are doing to one another; the De-scription, usually [performed] by the analyst, is the opposite movement of the In-scription by the engineer, inventor, manufacturer, or designer..."[35] Because of this reciprocal characteristic, cripting, and its family of semiotic tools have been embraced by Sociologists and designers alike.[36]

Through Scripting, designers may trace how their technologies will be understood and used step by step, and in the process, not only better understand the semiotic cues contained within their designs which create meaning for their users, but modify them through design. Ultimately, the designer *may configure the user*[37] through expectations, training, and/or Affordances (see the next section), then use the *script* to propel their users through a specific sequenced engagement with the artifact. In order for this to work, this script must be embedded within the artifact itself during the Design Process (Affordances describe how this may be accomplished). As Bruno Latour demonstrates in one of the more legendary examples of De-scription,[38] the Berlin Key, technologies can be designed to structure heterogenous relationships between human and non-human actants—in this case, not dis-

---

[34] Akrich, M. & Latour, B., 259–264.

[35] *Ibid*, 259.

[36] This is evident in a growing number of publications about scripting within the Design and allied disciplines, such as Kjetil Fallon's 2008 "De-scribing Design: Appropriating Script Analysis to Design History" in *Design Issues* 24(4): 61–75, or even further afield such as in Jørgensen, Finn Arne's 2013 "The Backbone of Everyday Environmentalism: Cultural Scripting and Technological Systems." In D. Jørgensen, F.A. Jørgensen & S. Pritchard (Eds.). *New Natures: Joining Environmental History with Science and Technology Studies*. University of Pittsburgh Press, 69–84.

[37] Woolgar, S. (1991). "Configuring the User: The Case of Usability Trials." In J. Law (ed.). *A Sociology of Monsters: Essays on Power, Technology and Domination*. Routledge, 57–99. and Grint, K. & Woolgar, S. (1997). *The Machine at Work: Technology, Work, and Organization*. Polity Press.

[38] Latour, B. (1992). "Where are the Missing Masses? A Sociology of a Few Mundane Artifacts." In W. Bijker & J. Law (Eds.). *Shaping Technology, Building Society: Studies in Sociotechnical Change*. MIT Press: 252–254.

similar to the way that Winner claims bridges may control access by 'gatekeeping.' The shape of the Berlin Key, in combination with the lock mechanism embedded within the door, set a scripted interaction into motion between a tenant and a concierge. The concierge may restrict a tenant's use of the key to the hours of their choosing by modifying the aperture of the keyhole. Typically, this can be choreographed such that the tenant's door must be locked at night but cannot be locked during the day. Hence, the design not only allows for a multiplicity of use but clearly places the determination of use (read as power) asymmetrically in the hands of the concierge. This example clearly illustrates how Scripting/De-scripting may construct and reveal not only technical aspects of how an assemblage works but also social values such as ethics, and agreements such as power dynamics, while reinforcing cultural roles and identities.

3.3.2 Affordances

Affordance theory may have originated in Jakob von Uexküll's 1909 notion of *Umwelt*, an biosemiotic-cybernetic theory that conceptualizes how organisms perceive their environment as a surrounding phenomenal-world that conditions their engagement with it.[39] However, the theory of Affordances wasn't codified until the late 1970s by Psychologist James J. Gibson's work on visual perception.[40] In turn, Gibson's theory was furthered by Design Scholar Don Norman in the 1980s, largely in relation to the emerging design field of computational interfaces.[41] Norman sees Affordances as relationships between objects and users where the objects themselves communicate to the user how they should be used.

A classic example that illustrates the power of a designed Affordance is a comparison between a standard teapot versus one where the handle is rotated 180° around the body of the pot, such that it is positioned under the spigot. In both cases, the handle communicates where and how the teapot should be held. In the case of the standard teapot, the user can use the handle to pour the hot tea neatly into a teacup without burning their hand. The teapot communicates to the user how to use it safely and effectively via the shape of the handle, and its placement. However, with one modification of the design, the unfortunate relocation of the handle onto the 'other side' of the teapot, disaster likely ensues if the user agrees to attempt to pour tea at all. Ironically, this version of the teapot also leverages Affordances, but in this case to communicate to users who have some prior familiarity with how to use teapots, not to attempt use this teapot in a normative way. It is a classic case of a materialized artifact that can *do more things with words/ ideas than things with the thing itself.*

---

[39] Von Uexküll, J. (2010 (1909)). *A Foray into the Worlds of Animals and Humans with A Theory of Meaning*. Translated by Joseph D. O'Neill. University of Minnesota Press.

[40] Gibson, J. J. (2015 (1979)). *The Ecological Approach to Visual Perception*. Psychology Press.

[41] Norman, D. A. (1988). *The Psychology of Everyday Things*. Doubleday.

For Norman, computational interfaces presented a new challenge for designers because they are primarily not physical. Hence, computational interfaces don't contain material and formal 'cues' that users of the time were accustomed to as guides through their effective use. These 'cues,' or Affordances, are not only critical for users to be able to intuitively navigate an artifact, interface, or system without a thick user manual, but in the case of the limited keyboard/monitor interface, needed to be entirely designed within the virtual reality of the software itself. Prior to Akrich's and Latour's work on Scripting, Norman was developing many of the 'scripting tools' to generate performative scripts, translations, and inscriptions as part of the construction of designed Affordances.

Though Winner never directly referenced the concept of Affordances in his unpacking of relationships between technologies and power (as far as this author is aware), they are everywhere between the lines in his work. A Design Affordance, as seen through Winner's analysis, is the mechanism by which agency can be designed into artifacts and their infrastructural systems and may continually condition/ govern the users of these artifacts/ systems. For example, Winner might argue that Moses was imagining the sequential script of buses picking up Manhattan passengers and beginning the drive to Long Island towards Jones Beach as a tool to understand where and how along their journey he might have the power to prevent them from reaching their final destination.

The functionality of both Winner's Politics of Design and Politics of Technological Systems rely on the ways that Affordances may be designed into artifacts and infrastructures to limit how users may engage with them, and as Winner ultimately argues, to exert power over them. Not dissimilarly, this chapter argues that Affordances can be used to create epistemic bridges between tools of Scripting and Affordances to connect scales, temporalities, and perspectives of designers grappling with predictive design approaches while managing the larger implications of their interventions. Likewise, these tools may be leveraged by more expansive collectives grappling with the political effects of existing designed artifacts and environments that shape our shared world. These all come together within the framework of System Oriented Design.

3.3.3 Systems Oriented Design

Systems Oriented Design [SOD] emerged in the late twentieth and early twenty first century, in part, to capture and leverage increasingly entangled complexities within contemporary systems infrastructures, and the Anthropocene. Developed across several contexts including computational-logistics design and management, numerous applications of game theory, and infrastructural architectures, among others, this chapter introduces its development as a tool largely tuned to the disciplines of Architecture and Environmental Design. Honed within the context of the Oslo School of Architecture and Design, and championed by Birger Sevaldson and Peter Jones, SOD furthers Scripting and Affordance approaches into relational configurations through the process of "Gigamapping," where complex situations and circumstances can be scoped, analyzed, and tested, in tandem with more traditional Design Processes.

In Sevaldson's 2022 book *Designing Complexity: The Methodology and Practice of Systems Oriented Design*,[42] he argues that Systems Thinking brings as much to the Design Disciplines as Design brings to Systems Thinking, and that ultimately, today's complexity requires the development of approaches that bridge across disciplinary knowledges and praxis through relational approaches. *Designing Complexity* introduces practices and methodologies that include extensive approaches to relational mapping, along with SOD Toolboxes and Process Frameworks formatted for use by designers. Like Scripting, SOD approaches can be 'reverse engineered' as tools of analysis for both scholars and designers. Sevaldson intentionally blurs the line between the two groups by encouraging both to use SOD to engage in frequent analysis, ideation, evaluation, and ultimately generation to create better collectives that serve our world in more democratic and effective ways. Because of this underlying SOD ethic, and its alignment with Winner's ultimate goal, SOD is a particularly potent design-analysis approach with which to carry Winner's project forward.

3.4 Conclusion

Despite being polemical in many ways (particularly for scholars of the Social Sciences), Winner's "Do Artifacts Have Politics" has had enormous impact across a range of disciplines, clearly withstanding the test of time. Generally, it is a fantastic introductory text to agency-driven relationships between the political and the material. Winner's clarity and didactic approach have consistently inspired a diversity of uses and have continued to fuel novel debates. When seen through the Design Disciplines, where practitioners have had very different kinds of exposure to social formations, Winner's ideas, illustrated in a straightforward manner, can be transformative. His premise is both empowering, and problematic for Designers in ways that complicate their more traditional practices, processes, and even identities, which by extension also apply to broader cultural collectives. Today, as the Design of technologies becomes increasingly complex, and reverberates through our equally increasingly complex societies via varied and surprising forms of agency, his work has proven more consequential than ever.

Acknowledgements The author is grateful to Mark Cabrinha and Barbara Allen for their interest, offerings, and support, which significantly enriched this chapter's treatment of Winner's background and context. She also extends a special thanks to the volume editors, particularly Alice Fox, for her numerous contributions above and beyond the norm.

[42] Sevaldson, B. (2022). *Designing Complexity: The Methodology and Practice of Systems Oriented Design*. Common Ground.

References

Akrich, M. (1992). The de-scription of technical objects. In W. Bijker & J. Law (Eds.), *Shaping technology, building society: Studies in sociotechnical change* (pp. 205–224). MIT Press.

Akrich, M., & Latour, B. (1992). A summary of a convenient vocabulary for the semiotics of human and nonhuman assemblies. In W. Bijker & J. Law (Eds.), *Shaping technology, building society: Studies in sociotechnical change* (pp. 259–264). MIT Press.

Barad, K. (2007). *Meeting the universe halfway: Quantum physics and the entanglement of matter and meaning*. Duke University Press.

Bennett, J. (2010). *Vibrant matter: A political ecology of things*. Duke University Press.

Bijker, W. (1987). The social construction of Bakelite: Toward a theory of invention. In W. E. Bijker, T. P. Hughes, & T. Pinch (Eds.), *The social construction of technological systems* (pp. 155–182). MIT Press.

Bijker, W., Hughes, T. P., & Pinch, T. (Eds.). (1987). *The social construction of technological systems*. The MIT Press.

Brand, S. (1994). *How buildings learn: What happens after they're built* (p. 35). Penguin Books.

Busbea, L. D. (2021). *Proxemics and the architecture of social interaction*. Columbia University Press: Columbia Books on Architecture and the City.

Davis, J. (2020). *How artifacts afford: The power and politics of everyday things*. MIT Press.

Davis, J., & Chouinard, J. (2016). Theorizing affordances: From request to refuse. *Bulletin of Science, Technology & Society, 36*(4), 241–248.

Ellenzweig, S., & Zammito, J. (Eds.). (2017). *The new politics of materialism: History, philosophy, science*. Routledge.

Fallan, K. (2008). De-scribing design: Appropriating script analysis to design history. *Design Issues,* 24(4), 61–75.

Foucault, M. (2007). *Security, territory, population: Lectures at the Collège de France, 1977-1978* (M. Senellart, Ed.; G. Burchell, Trans.). Palgrave Macmillan.

Gibson, J. J. (1982). *Reasons for realism: Selected essays of James J. Gibson*. E. Reed & R. Jones (Eds.). Lawrence Erlbaum.

Gibson, J. J. (1997). The theory of affordances. In R. Shaw & J. Bransford (Eds.), *Perceiving, acting, and knowing: Toward an ecological psychology* (pp. 67–82). Lawrence Erlbaum.

Gibson, J. J. (2015 [1979]). *The ecological approach to visual perception*. Psychology Press.

Griffero, T. (2020). Architectural affordances: The atmospheric authority of spaces. In *Places, affordances, atmospheres: A pathic aesthetics* (pp. 99–114). Routledge.

Grint, K., & Woolgar, S. (1997). *The machine at work: Technology, work, and organization*. Polity Press.

Gross, M. (2010). *Ignorance and surprise: Science, society, and ecological design*. MIT Press.

Gross, M., & McGoey, L. (2015). *Routledge international handbook of ignorance studies*. Routledge.

Halprin, L. (1970). *The RSVP cycles: Creative processes in the human environment*. G. Braziller.

Haraway, D. (1985). Manifesto for cyborgs: Science, technology, and socialist feminism in the 1980's. *Socialist Review,* 80, 65–108.

Haraway, D. (2003). *The companion species manifesto: Dogs, people, and significant otherness*. Prickly Paradigm Press.

Harley, A. (2019). Haraway's material-semiotic knot: A learning-teaching response for creative-critical times. *a|b: Auto|Biography Studies, 34*(3), 539–553.

Heft, H. (1989). Affordances and the body: An Intentional analysis of Gibson's ecological approach to visual perception. *Journal for the Theory of Social Behaviour, 19*(1), 1–30.

Hess, D. (2016). *Undone science: Social movements, mobilized publics, and industrial transitions*. MIT Press.

Hughes, T. (1983). *Networks of power: Electrification of Western Society, 1880-1930*. John Hopkins University Press.

Jasanoff, S. (Ed.). (2004). *States of knowledge: The co-production of science and the social order*. Routledge.

Joerges, B. (1999a). Do politics have artifacts? *Social Studies of Science, 29*(3), 411–431.

Joerges, B. (1999b). Scams cannot be busted. *Social Studies of Science, 29*(3), 450–457.

Jørgensen, F. A. (2013). The backbone of everyday environmentalism: Cultural scripting and technological systems. In D. Jørgensen, F. A. Jørgensen, & S. Pritchard (Eds.), *New natures: Joining environmental history with science and technology studies* (pp. 69–84). University of Pittsburgh Press.

Kaptelinin, V. (2014). *Affordances and design*. The Interaction Design Foundation.

Klein, H., & Kleinman, D. L. (2002). The social construction of technology: Structural considerations. *Science, Technology, & Human Values, 27*(1), 28–52.

Latour, B. (1992). Where are the missing masses? A sociology of a few mundane artifacts. In W. Bijker & J. Law (Eds.), *Shaping technology, building society: Studies in sociotechnical change* (pp. 252–254). MIT Press.

Latour, B. (2005). *Reassembling the social: An introduction to actor-network-theory*. Oxford University Press.

Law, J. (2023). Material semiotics. In P. Atkinson, S. Delamont, A. Cernat, J. W. Sakshaug, & R. A. Williams (Eds.), *SAGE research methods foundations*. Sage.

Lemke, T. (2015). New materialisms: Foucault and the 'government of things'. *Theory, Culture & Society, 32*(4), 3–25.

Lemke, T. (2021). *The government of things: Foucault and the new materialisms*. New York University Press.

McGrenere, J., & Ho, W. (2000). Affordances: Clarifying and evolving a concept. *Graphics Interface*, 179–186.

Museum of Modern Art [MOMA]. (2025). Readymade. Retrieved March 18, 2025, from https://www.moma.org/collection/terms/readymade

Noble, D. (1984). *Forces of production: A social history of industrial automation*. Alfred A. Knopf.

Norman, D. A. (1988). *The psychology of everyday things*. Doubleday.

Norman, D. A. (1999). Affordance, conventions, and design. *Interactions, 6*, 38–42.

Norman, D. A. (2013). *The design of everyday things, revised and expanded edition*. Basic Books.

Rittel, H., & Webber, M. (1973). Dilemmas in a general theory of planning. *Policy Sciences, 4*(2), 155–169.

Sanders, J. T. (1993). Merleau-Ponty, Gibson, and the materiality of meaning. *Man and World, 26*, 287–302.

Sanders, J. T. (1997). An ontology of affordances. *Ecological Psychology, 9*(1), 97–112.

Sanders, J. T. (1999). Affordances: An ecological approach to first philosophy. In G. Weiss & H. F. Haber (Eds.), *Perspectives on embodiment: The intersections of nature and culture*. Routledge.

Sevaldson, B. (2022). *Designing complexity: The methodology and practice of systems oriented design*. Common Ground.

Suchman, L. (1985). *Plans and situated actions: The problem of human-machine communications*. Xerox.

Suchman, L. (2007). *Human–machine reconfigurations: Plans and situated actions* (2nd ed.). Cambridge University Press.

Turner, F. (2006). *From counterculture to cyberculture: Stewart Brand, The whole earth network, and the rise of digital utopianism*. University of Chicago Press.

Von Uexküll, J. (2010 [1909]). *A foray into the worlds of animals and humans with a theory of meaning* (J. D. O'Neill, Trans.). University of Minnesota Press.

Winner, L. (1980). Do artifacts have politics? *Daedalus, 109*(1), 121–136.

Winner, L. (1986). *The whale and the reactor: A search for limits in an age of high technology*. University of Chicago Press.

Winner, L. (2004). *Technology and social theory*. Syllabus, Fall, Rensselaer Polytechnic Institute.

Woolgar, S. (1990). Configuring the user: The case of usability trials. *The Sociological Review, 38*(1), 58–99.
Woolgar, S. (1991). Configuring the user: The case of usability trials. In J. Law (Ed.), *A sociology of monsters: Essays on power, technology and domination* (pp. 57–99). Routledge.
Woolgar, S., & Cooper, G. (1999). Do artifacts have ambivalence? *Social Studies of Science, 29*(3), 433–449.

Chapter 4
"Do Artifacts Have Politics?": From Material to Immaterial Dimensions of Urban Infrastructures

Federica Cagnoli

> *The issues that divide or unite people in society are settled not only in the institutions and practices of politics proper, but also, and less obviously, in tangible arrangements of steel and concrete, wires and transistors, nuts and bolts.*
>
> — *Winner (1980, p. 128)*

Abstract Urban infrastructures are known to determine the quality of a city, such as its justness and democratic approaches. They enable the flow of goods, people, and information. More importantly, thanks to them, people can develop a new, or different, understanding of their own environment and themselves, both thanks to their personal standpoints and encountering others. However, even though many scholars have recognized such a crucial role, a theoretical review of this theme and its broader implications seems to be missing.

In this chapter, the author intends to fill this gap by drawing on Science and Technology Studies and the work of Langdon Winner. She will focus on pieces of research dealing with urban infrastructures and their relationship with cities, and on how this can influence the daily life of people. In doing so, she will present urban infrastructures according to what she calls 'double dimension': Material and immaterial. In this sense, one will be able to focus on the social implications created by these massive projects. Indeed, given their acknowledged role in influencing cities,

A variation of this chapter was previously published in Interactions between urban infrastructures and the liveability of cities: a comprehensive review. In J. Montgomery (ed) AMPS Proceedings Series 34.3, Liveable Cities: A Conference on Issues Affecting Life in Cities, pp. 47–54.

F. Cagnoli (✉)
Department of Political and International Sciences, University of Genoa, Genoa, Italy
e-mail: federica.cagnoli@edu.unige.it

© The Author(s), under exclusive license to Springer Nature Switzerland AG 2026

A. Fox, D. Cressman (eds.), *Arrangements of Power: Tracing Langdon Winner's Legacy Within and Beyond the Philosophy of Technology*, Philosophy of Engineering and Technology 50, https://doi.org/10.1007/978-3-032-17371-3_4

it is necessary to consider how urban infrastructures can affect issues such as social integration, uneven development, and growing masculinity.

This chapter has a twofold goal. First, it can make a contribution to academic research by providing a summary of what the field has said so far on the topic, which can help scholars identify new elements to advance in research. Second, it can present practical hints. Thanks to the presentation of pivotal aspects in the double dimension of urban infrastructures and how these can impact and be impacted by cities, one can develop better practices for these projects.

Keywords Urban infrastructures · Science and technology studies · Urban development · City-infrastructure interactions · Social integration

4.1 The Context

This chapter pertains to the study of boring things, in other words, infrastructure, which often seems to be an infinite list of numbers and technical details (Star, 1999). However, spaces such as cities, schools, or work environments are deeply impacted by infrastructure. Ignoring these means ignoring fundamental aspects of our society, which range from justice to planning for power (Latour & Hermant, 1998). For this reason, it is necessary to identify new methods to better comprehend the overlap between infrastructures, in this case urban, such as highroads and railways, and human organisations (Star, 1999).

Considering the vital role they can play; it is apparent why many studies gave their own definition of urban infrastructures. Based on the extensive literature in Science and Technology Studies (STS) and philosophy of technology (the latter intended to understand technologies as "political phenomena in their own right", Winner, 1986 p. 22) *that, starting in the 2000s, has seen a growing importance of the topic, the following passages present the different definitions given, emphasising shared or peculiar aspects.*

Steele and Legacy (2017) describe infrastructure as a multidimensional phenomenon that has profound social, political, and economic implications, which is as much about space, ecology, and culture as it is about pipes, scaffolding, cables, and concrete. In the Global North, infrastructure systems are generally viewed as networks that facilitate the flow of people, goods, energy, water, waste, and information (Addie et al., 2020). In material terms, Siemiatycki et al. (2020) defines infrastructure as a vast physical and technical network of roads, tunnels, railways, public transport networks, and much more, which are installed to satisfy society's growing desire for connectivity, communication, and exchange. Focusing on megaprojects, especially the transnational ones, these can also be defined as large-scale projects, implemented in a collaborative way between two or more states, that intentionally and profoundly transform landscapes and require a coordinated management of capital and state power (Ariemma & Burnside-Lawry, 2016; Gellert & Lynch, 2003). More concisely, Frick (2014: 22) states that infrastructure megaprojects must be characterised by the six 'C's': 'Colossal, captivating, costly, controversial, complex, and subject to issues of control'. As will be explained in this chapter, although they differ in some respect, these definitions and characteristics share a social dimension that is influenced by and influences the infrastructure.[1]

In this sense, this chapter does not want to give a restrictive definition of urban infrastructures. The richness and diversity of the phenomenon would be limited. The purpose is to theoretically present the potential that infrastructures can have in cities. Indeed, urban infrastructures, which can be considered as technologies, can create order in our life and, at the same time, impact it: They can influence how people communicate, travel, and much more. In this sense, one may ask what the implications of different designs of infrastructures can be on ordinary life (Winner, 1979).

This chapter intends to present how urban infrastructures can affect cities and vice versa, such as the experiences of citizens, by drawing on STS and philosophy of technology, paying particular attention to the work of the political theorist Langdon Winner. Doing so means being aware that urban infrastructures are not neutral (Winner, 1979) and thus have implications that cannot be ignored.

The chapter will present urban infrastructures according to what the author calls the 'double dimension': Material and immaterial. This perspective will enable one to see how, over decades, urban infrastructures have impacted and have been impacted by cities, focusing specifically on the work undertaken at the social level of the former. Indeed, given the crucial role they play, it is necessary to consider ethical dilemmas going beyond the mere analysis of impacts and side effects of these projects (Winner, 1983) and looking at the broader consequences for human well-being and social justice brought by technological advancements.

Bearing in mind that current social processes affect and are affected by the interactions between scientific knowledge, technological advancements, and corporate interests, this chapter has a twofold goal. On the one hand, it contributes to academic research by providing a presentation that can help scholars identify new elements to advance research. On the other hand, it presents practical implications: By presenting key components in the relationship between urban infrastructures and cities, it can pave the way toward better practices in planning, building, and maintaining these projects.

[1] For a detailed analysis of the tension between technological determinism (the idea that technology shapes society) and social constructivism (the idea that society shapes technology), see, for instance, Winner (1986).

4.2 The Theoretical Framework

Urban infrastructures can be studied in different ways. For instance, one may wonder whether they are technological systems, financial instruments, biological[2] (as being embodied in the physicality of humans), or social (as being composed of diverse human practices) (Zaloom, 2003). In STS, they can be situated into three 'levels': The level of individuals and small groups, the level of institutions (such as corporations and standard bodies), and the level of political economies and government (Edwards, 2003). By employing STS perspectives and the works of Winner, infrastructures are considered here as practical artifice (Winner, 1980), which can become networks that enable the movement of other matter (Larkin, 2013). They are a material bundles of relationships that are created and dismantled by humans and other actors. They are things and, at the same time, the relation among things (Larkin, 2013). In this sense, such bundles not only work in the abstract, but they can also favour the creation of contestation, inequalities, and the production of the state and its power: In other words, they are ways of life and a constitution of a sociotechnical order (Winner, 1979, 1983).

STS scholars have analysed how these infrastructures can mediate exchange over distance to enable the interaction among people, objects, and spaces, and therefore the base for the modern social system (Larkin, 2013). Moreover, STS is grounded in the constructivist perspective, which emphasises that technological and social systems are co-produced rather than developing independently. For example, Actor--Network Theory (e.g., Latour, 2007) highlights the role of non-human agency. The alignment of people, technology, and resources is equally important (Matthewman & Byrd, 2020). This perspective helps illuminate the connections between humans and urban infrastructures.

This means that urban infrastructures are not neutral. They can embody specific forms of power and authority (Winner, 1980, 1986), and, in this sense, one has to ask what gives them that specific shape (Winner, 1979). For instance, one can think of the impact of the industrial revolution on society as a whole, which then competed with political institutions to see who could have power and authority over people. As such, infrastructures are privileged actors to question the assumption according to which technology and society stand in opposite ways.

Given the flows of elements such as information, cultural products and people, social and cultural phenomena cannot be considered isolated and studied as a unique site (Silvast & Virtanen, 2019). As infrastructures can be matched to types of society (Deleuze, 2011), thus being democratising and liberating forces, or the opposite, they are critical sites of deliberation about what and whose social and economic standpoints are supported (Carse, 2016).

However, one should not argue that urban infrastructures have political properties by themselves because they are not neutral. When studying them, it is important to take into account the social circumstances in which they are developed, built, and

[2] To delve into the idea of technologies that evolve and adapt as biological entities, see Winner (1983).

used, as well as their features. In this sense, one should ask if adopting a certain technology needs the creation of a specific social conditions (Winner, 1980).

According to Winner (1980), there are two ways in which artifacts—here, urban infrastructures—have political features. First, the infrastructure becomes a way to settle an issue that is lived by a given community. Second, certain infrastructures are inherently political, as they are linked to particular political relationships, the latter meant as arrangements of power and authority. When this is considered, as already mentioned, such infrastructures influence how people live their ordinary lives, and their differences in power and awareness play a role in the level of influence exerted.

In view of what has been explained, specific aspects are considered here to better capture the aim of this chapter. STS has tended to privilege technical aspects and overlooked less refined approaches to economic, social, and political processes (Furlong, 2011). In particular, it is necessary to look at the social consequences of technologies, which means questioning what relevant social groups and social interests are, and identifying dynamics behind involved social groups and actors (Winner, 1993). Similarly, there is less academic debate about the relation between infrastructures and the patterns of practice they sustain (Shove, 2015). Thus, considering the following elements can provide a more comprehensive overview of urban infrastructures. It is important to look at the social consequences of technical choices; in other words, what is shaping the technology so that it is producing that specific effect, and what are the causes of the effects that people are living. Another aspect to be aware of is deciding the relevance of groups, especially those without a voice, and interests at stake when dealing with urban infrastructures. This fact also means that there may be other dynamics behind the needs, interests, problems, and solutions of groups and social actors. All in all, doing so means going beyond the so--called value neutrality. Even though studying how some people see an infrastructure in one way while others see it differently can help reveal demands and needs in a technical development, a further step is needed. In a world where profound controversies revolve around designs and uses of infrastructures, one has to evaluate the features of such controversies, especially the many points of view influencing choices on technology.

Overall, this theoretical framework encourages the consideration of the broader picture, not only the technological aspects of urban infrastructures but also its networks, both those that support it and those that it sustains. Doing so goes beyond things-in-themselves, reaching the so-called "relationships-in-the-making" (Kasdogan, 2017, p. 24).

This chapter focuses on what infrastructures create in cities for people's ordinary life and activities. The author looks at a peculiar feature of urban infrastructures, which she contrasts as material and immaterial. Then, the remaining themes explored in this chapter are included in Sect. 4.4, and are Sects. 4.4.1, 4.4.2, and 4.4.3.

4.3 The Double Dimension of Infrastructure

As the paragraphs described above clearly state, infrastructures must be considered to have a double dimension. In this study, this is indicated as *material* and *immaterial*. Although the following two sections delve into studies and research conducted on these two elements separately; to study the infrastructure, it is necessary to read them as if they overlapped. In this way, it will be possible to have a clear overview of the context.

4.3.1 The Material Dimension

Over the centuries infrastructure has been increasingly used to transport goods and services, such as in high-speed lines. Therefore, these projects are precisely created to change social structures (Flyvbjerg, 2014). By making possible the creation of a global system of transportation, communication, and logistics, infrastructures have facilitated what some refer to as the 'Great Acceleration': A dramatic increase in population, economic activity, and resource use to maintain the infrastructure itself (Hibbard et al., 2006). Even today, infrastructures remain central to political agendas worldwide, so much so that, in 2008, the Economist (Anon, 2008) considered infrastructure spending the most extensive investment in human history. Recently, Dodson (2017) spoke of an international infrastructural turnaround. The resulting investments enable the creation of jobs and the maintenance of the economy moving. If well idealised, infrastructures can positively impact many other aspects of society, such as maintaining high employment levels or improving productivity and competitiveness. On the other hand, if the investment is not well-conceived, the risk is weakening the economy. However, economic development is not the only aspect to consider in order to explain infrastructural growth: Indeed, this can be seen as a symbol of prosperity, vitality, and political leadership (Siemiatycki et al., 2020).

Despite the role played by infrastructures, much information shows that they rarely manage to achieve the expected results within the preestablished period and according to the budget, so much so that it is said that only one project out of ten succeeds in each of these aspects (Cantoni & Pagnone, 2020). In light of such findings, Flyvbjerg (2014, p. 11) summarises this phenomenon through the 'iron law': Over budget, over time, and over and over again.

Taking into account such a vital stake and the different management of infrastructures, Flyvbjerg explains the boom of megaprojects and infrastructures through the 'sublime four' (2014, p. 8): Technological, political, economic, and aesthetic. Regarding the former, the focus is on the experts and their willingness to create ever more innovative projects, overcoming the limits of what technology can do. The second, political, is about interests that politicians have in building projects for themselves and their causes, with growing visibility that can help them gain votes during the elections. Continuing with the economy, actors in business and trade, as well as unions, can obtain new jobs thanks to infrastructures. Finally, aesthetics refers to projects that can also be beautiful, such as the Golden Gate Bridge in San Francisco. In summary, infrastructure megaprojects are not only a representation of modern civilisation but are also capable of generating feelings of wonder and awe towards them and their builders. Such a sublimity does not represent a problem in itself. However, what must be taken into account is the 'insatiable ego' that lies behind many of these projects (Flyvbjerg, 2017). To this fact, one must link what Kaika and Swyngedouw (2000) call the 'phantasmagoria of urban technical networks', which often translates into a lack of attention to projects that are less spectacular but respond directly to primary social problems. The result of these processes is what Flyvbjerg (2009) calls 'the survival of the unfitted': Unlike Darwin's law, among infrastructures, the worst projects tend to survive because the benefits are exaggerated, while the costs are underestimated. Many cases support this thesis, such as the Eurotunnel, which connects England and France passing under the English Channel, or the connection of the Great Belt in Denmark. In both cases, the costs exceeded what was initially decided. The projects were only completed because the cost-benefit ratio presented to investors was considerably inflated (Flyvbjerg, 2009).

In other words, one cannot consider the material aspect as independent of political and organisational facets, financial needs, and know-how necessary for its design, construction, operation, and maintenance (Filion & Keil, 2017). For these reasons, it is also essential to understand the power relations that come into play in all phases of project development, from agenda setting to financing. Consequently, as will be explained, this production cannot be seen as an aspect solely related to engineering: There are informal processes of creation, use, and adaptation that make the infrastructure a political project of production and communication of information (Siemiatycki et al., 2020).

Infrastructures symbolise modernity and scientific progress, representing alternative futures ranging from governance based on efficiency and resilience to economic failures and collapses (Addie et al., 2020). However, the infrastructure does not often operate as established. This fact does not concern only the economic impact. In fact, even when development progresses as planned, there can be negative impacts on gender, class, and ethnicity. For this reason, it is necessary to emphasise the immaterial dimension that characterises urban infrastructures.

4.3.2 The Immaterial Dimension

As underlined above, the social aspect is fundamental in infrastructures. It is possible to see the infrastructure as a lens for understanding socio-spatial and political dynamics and as a technology that catalyses and crystallises these dynamics (Addie et al., 2020).

That is, since the infrastructure exists in space and, at the same time, creates space thanks to its technical functions (Easterling, 2014), people define and experience their environment also thanks to its presence. For these reasons, although infrastructures can be interpreted as distinct objects of analysis, in this study particular attention is paid to the relationships they create. This point of view becomes even more pertinent if one looks at the work that infrastructure does in the social field. In this sense, Calhoun (1992) states that the large-scale social relations of the contemporary era are principally made possible by infrastructures. Think, for example, of the meetings one has while taking the bus. Thanks to these, people can develop a perception of the environments in which they live, expectations, battles to fight, and consequent changes on the basis of their power and position in the infrastructural system (Angelo & Hentschel, 2015). As a result of the multitude of interactions across and in infrastructures, people can develop patterns of social practices, understandings, and relationships. Following this perspective, one can consider, for example, the transformations created by the subway in New York in 1904. By leading people to a new social role and bringing them closer, the subway produced a new subject: The passenger (Angelo & Hentschel, 2015). Designers were interested in bringing people together through public transportation, which improved the circulation in the city.

However, the effects on people are not always positive. Staying in New York, among the most famous pieces by Winner (1980), there is the case of the bridges over the Long Island Parkway. Robert Moses, the city planner, decided to build low bridges to prevent public buses from passing under them and reaching the wealthier suburbs of Long Island. The limitations, in this case, were not due to the will of politics but to the designers. In the same years as the New York subway, Cairo and Berlin developed urban railway lines that instead of bringing people together, drove them away (Prestel, 2015). The reference group in these two cases was the emerging middle class in order to allow them to live a healthy and happy life in suburbs far from the two cities. Unlike the New York subway, these two railroads fragmented people's experience, making it easier and faster for the wealthy to move from the chaotic city centre and reform their bodies and minds in their new dwellings. In conclusion, there are millions of infrastructural bridges and just as many buses that cannot pass under them (Star, 1999). Each potential change is inscribed profoundly in the design of these projects: Some are more malleable than others, while others need external help, such as social movements, to be accomplished.

As was said at the beginning of the chapter, infrastructures are not just things but bundles of relationships. Individuals and communities gather around them for collaborations, deliberations, or in conflict, creating implications of different kinds (Carse, 2016; Star, 1999). In this sense, infrastructures enable one to understand change: By materially changing people's daily lives, one can experience large-scale social transformations. Although it is difficult to see these transformations when one experiences them, through interactions with infrastructures and with other individuals, one can obtain knowledge about what is happening and develop a sense of oneself as a resident of a city or a member of a nation (Angelo & Hentschel, 2015). However, this process is not necessarily automatic: The encounters are often opaque, since infrastructures are difficult to conceive. From this point of view, Eric Trovalla and Ulrika Trovalla (2015) define this understanding precisely as 'divining'. In their study of Jos, Nigeria, they show how it is possible to collect information thanks to the infrastructure. Yet, some studies focusing on Large Technical Systems (LTS), a discipline in the history of technology, underline the inertia of these systems (e.g., Hughes, 1994). According to these, the infrastructure does not grow from scratch or in a different way but struggles against the inertia of its own base, inheriting weaknesses and strengths of that base (Star, 1999).

In addition to durability, stability, and high costs, other factors make infrastructure strongly path dependency and against sudden changes. Returning to the double dimension, social groups, which intervene and participate in implementing and maintaining infrastructures, have interests and interactions equally significant (Monstadt, 2009). According to this perspective, since the material and immaterial elements are strongly intertwined, changing one component often means consequences for the entire system. For this reason, changes, although they can be massive and impact different areas of society, require time, negotiations with the actors concerned, and consequent adjustments (Star, 1999).

4.4 The Politics in Urban Infrastructures

The presented studies state that infrastructures shape and are shaped by cities. By being made of physical components, such as water pipes, power plants, or sewage networks, opportunities and constraints are created for social changes that otherwise could not occur. At the same time, infrastructures have social components, such as professional organisations, users, and social networks. These parties decide how to provide infrastructure services by building, managing, and changing these artefacts (Monstadt, 2009). This implies that infrastructures, as well as affecting the environment, impact bodies, behaviours, and entire populations.

4.4.1 Heralds of Meaning

Although, as explained, changes in infrastructures are a complex aspect to identify and control, the infrastructure carries and represents different meanings. One of the most evident examples concerns the symbolic messages that the promoters of these public projects want to spread.

Infrastructure construction cannot be conceived as the design of an object but also as a political project that creates and communicates something (Siemiatycki et al., 2020). To understand infrastructures, they must be examined not only for their technical functions but also as semiotic vehicles (Larkin, 2013). The ‘poetics of infrastructure’ of Larkin (2013, p. 329) refers precisely to the ways in which infrastructures transmit social and cultural content. To understand this semiotic capacity, its link with the local landscape must be considered. In this regard, Mukerji (2009), referring to the Eurotunnel, says that the latter has become a brutal fact in the countryside to go around like a mountain, adding that people see the infrastructure not as a network that crosses places, but as something intertwined with them. On the other hand, Carse (2014) points out that people living near the Panama Canal do not see the infrastructure as a technical system or a node of a broader global network, but as part of an ever-changing landscape, with rivers that become lakes and roads that are first built and then abandoned.

Very often, messages transmitted do not reflect what the infrastructure actually does: They concern issues such as the values of promoters, social objectives, or political power (Siemiatycki et al., 2020). This capability allows infrastructures to create a precise vision of society. This fact explains why they are often at the centre of debates and discussions: By being able to plan, communicate, and interpret this system, a group is able to support existing or new hierarchies of power (Siemiatycki et al., 2020).

The ability to connect different levels and people has implications that go far beyond achieving a world in motion. Thanks to the experiences of and in infrastructures, it is possible to study how social patterns, relationships, and urban imaginaries are structured. In other words, infrastructures allow us to see how knowledge is produced. Recent studies in the field focus on this issue, for example, to help understand the role of infrastructure in reinforcing or dismantling unsustainable behaviour (Carse, 2016). Based on this perspective, the following two studies by Taber (2016) and Jalbert (2016) can be considered. The first focuses on oil exploration in a national park in Ecuador, which eventually turns into a debate on the responsibility and visibility of oil extraction. The second, based on ethnographic participations and participatory action research, concerns civil society groups in the United States that are intent on producing new knowledge through tests on water quality. They aim to challenge a fossil fuel extraction network. In both cases, the authors want to study environmental problems related to infrastructures and how knowledge can make these problems more visible, thus raising awareness and allowing informed and democratic debates.

4.4.2 Producers of State and Power

In light of what has been traced in these pages, it seems clear that the infrastructure can also produce state and power thanks to assembling the social and changing the relationships between communities and states (Acevedo Guerrero, 2018). This chapter has already discussed how such spectacular public works can exalt the state and the politicians involved. In this regard, according to some scholars, since the 1990s, infrastructural investments no longer serve to modify the territory but to underline the role of the state and its capabilities: From this moment on, the word 'infrastructure' increasingly tends to refer to topics such as human capital or regulations (Hetherington & Campbell, 2015). This aspect is notably visible in Latin America, where the idea of infrastructure has begun to coincide with the model of industrialised society (Hetherington & Campbell, 2015). For example, Cohen

(2015) describes how, in Colombia, the infrastructures used to extract raw materials are constantly linked to the artificial environment (built by the state), violence, and the control of local female miners.

Taking up the concept of infrastructure as that of a political terrain in which power relations are created, reproduced, and contested, one can see infrastructures as composed of multiple levels that work on different aspects. They can connect or distance people, goods, flows, and ideas (Acevedo Guerrero, 2018). By studying their material components, it is possible not only to study the past of a society but also its future, since, as already described, infrastructures are also about people's aspirations and imagined futures. Many studies focused on these issues, especially on the relations between communities and states. Some of these date back to the 1970s and 1980s: Akrich (1992) focused on the political capabilities of infrastructures in the Ivory Coast. By examining the local electrification process, she shows its ability to reach hitherto isolated regions, shaping and reflecting social arrangements. The electricity grid creates a new relationship between technology, the state, and citizens, who are now self-sufficient individuals rather than collective actors. In doing so, the electricity grid reinforces the distinction between the private and public thanks to framing consumption as a private matter. More recently, Meehan (2014), by analyzing the water infrastructure in Tijuana, Mexico, documents its role in producing state power and limiting it. If, on the one hand, infrastructures help territorialize and consolidate federal authority over the years, on the other, the continuous use by citizens of household tools—such as barrels or cisterns—to collect rainwater prevents stabilization. From this point of view, when a community is unwilling or unable to fulfil the role that the infrastructure has given it, it is necessary to persuade people to behave in a certain way. According to Akrich (1992), persuasion is fundamental to achieving infrastructure stability: Without it, the infrastructure remains a chimaera. In this regard, studies in Managua, Nicaragua, and Barranquilla, Colombia, show the communities' opposition to the roles they were supposed to play in electricity and water infrastructures. These people challenge bills in court and government agencies and, in some cases, act materially, for example, by distorting consumption measurements (Acevedo Guerrero, 2018).

4.4.3 Developers of Cities

As already assessed, infrastructure, thanks to mediating resource flows, plays a crucial role in defining and modifying the environment and the impact it has on communities. Thanks to such an aspect, one can study the role infrastructures have in producing environmental problems and behaviors that further damage it, as well as solutions. Since infrastructures have meanings and need continuous maintenance, they can become a place for debates on the distributions of environmental costs and benefits, demonstrating how the infrastructure for one group can become the problem of another (Carse, 2016). In summary, these projects are inherently ambivalent: They can be seen as root causes of many environmental questions and fundamental in solving them (Monstadt, 2009). Nevertheless, it is necessary to consider infrastructures also from another perspective, that of the territory. Indeed, they represent a new field in which one can examine forms of spatial planning, particularly from a development viewpoint (e.g., Schindler & Kanai, 2021). The *Laboratoire Techniques, Territoires et Société* (Anon, n.d.-a) of Paris Est University deals with infrastructure policies, technologies, and territoriality. Similarly, 'Mistra Urban Futures' (Anon, n.d.-b) was a research program on sustainable urban development. Regarding infrastructure and territory themes, it mainly considered the former as a political and material terrain where one can create new cities.

Again, focusing on cities, infrastructures are a fundamental element from several points of view. Due to the massive impact that these have on economic and social development, the well-being of cities (and, consequently, of the states) can be influenced by one project, more than most when a failure follows it. For instance, as described by Flyvbjerg (2014), continuous delays for the inauguration of an international airport in Hong Kong forced air traffic to move to other areas, with a consequent fall in the GNP of the region. However, this bond goes beyond this aspect: Contemporary societies are increasingly urban. Consequently, cities become the pivotal places where infrastructure resource flows are geographically concentrated and where daily exchanges between the infrastructure and natural environments occur (Monstadt, 2009). As cities are expanding their network of action, importing resources and using neighboring regions, infrastructures become even more relevant to set these mechanisms in motion. In other words, it is possible to state that modern and contemporary development of cities has depended, and continues to do so, on infrastructure networks that guide the resource flows within and along cities (Monstadt, 2009). More precisely, according to the LTS framework, cities and infrastructures have evolved according to a dialectical relationship. On the one side, the development of cities has been made possible by infrastructures, which guide and facilitate urban life and functions in different ways (Monstadt, 2009). On the other one, the urban environment is still today a precondition for the development, innovation, and growth of these projects (Coutard et al., 2005). Despite the significant link in which the dependence of cities on the functioning of these works becomes clear, as already discussed, infrastructures often become invisible and taken for granted: They are ignored unless they result in an apparent failure (Monstadt, 2009).

Today, such dynamics are predominantly visible in the southern hemisphere, which sees massive infrastructure investment in order to accelerate urbanisation and connect neighbouring cities. According to some scholars, this phenomenon is connected to the infrastructural drive. According to others, however, focusing on the global dimension means ignoring an equally considerable parallel process: State urbanisation. Indeed, in countries where capital accumulation is mainly concentrated in large conurbations, the policy has been defined over the years by state-level efforts to implement these projects and by the grievances surrounding subsequent infrastructures and urban developments (Shatkin, 2022).

4.5 Conclusion

The overview presented here shows, according to the double dimension of urban infrastructures, the capacity of these projects to carry out work in social terms. The interactions occurring within and with infrastructures are the foundations of social forms that can be on a higher level of location. Such social models include patterns of social integration—or fragmentation—as seen for the subway in New York or the railway lines in Berlin and Cairo. Then, there is the work of producing knowledge to make visible, for example, environmental problems and their consequent debates, as well as the production and limitation of state power, as the water infrastructure case in Mexico makes clear. Finally, among the principal works retraced in these pages, there is the (unequal) economic and geographical development, which can bring benefits and costs within the territory, depending on how the project is conceived. In this regard, in recent decades, several studies have documented, according to a gender and ethnic perspective, the unequal results brought by infrastructures or the different experiences that people undergo when using these (e.g., Urry, 2006). In order to respond to these diverse effects, some scholars have questioned the material and symbolic Eurocentrism that characterizes these works and adopted a postcolonial and feminist point of view, studying how infrastructures are built, where and by whom (Siemiatycki et al., 2020).

Infrastructures, being at the same time an influence for and influenced by cities, can affect the latter and, therefore, people's ordinary life and activities. Being aware of this aspect can help advance research. However, as this chapter has demonstrated, this is not enough. The double dimension of infrastructures has demonstrated the need to recognise their lack of neutrality (Winner, 1979). As a result, one has to look at what gives infrastructures their peculiar shape, taking into account the social circumstances, including consequences and groups involved, in which urban infrastructures occur (Winner, 1979). By using such an understanding, we may be able to answer how we can build and investigate urban infrastructures that can make society free, just, and sustainable.

References

Acevedo Guerrero, T. (2018). Water infrastructure: A terrain for studying nonhuman agency, power relations, and socioeconomic change. *WIREs Water, 5*(5). https://doi.org/10.1002/wat2.1298

Addie, J.-P. D., Glass, M. R., & Nelles, J. (2020). Regionalizing the infrastructure turn: A research agenda. *Regional Studies, Regional Science, 7*(1), 10–26. https://doi.org/10.1080/21681376.2019.1701543

Akrich, M. (1992). The de-scription of technical objects. In W. E. Bijker & J. Law (Eds.), *Shaping technology/building society: Studies in sociotechnical change* (pp. 205–224). MIT Press.

Angelo, H., & Hentschel, C. (2015). Interactions with infrastructure as windows into social worlds: A method for critical urban studies: Introduction. *City, 19*(2–3), 306–312. https://doi.org/10.1080/13604813.2015.1015275

Anon. (2008, June 7). Building BRICs of growth. *The Economist,* 80

Anon. (n.d.-a). LATTS—Laboratoire Techniques, Territoires et Sociétés. Retrieved November 20, 2022, from https://latts.fr/

Anon. (n.d.-b). Mistra Urban Futures. Retrieved November 20, 2022, from https://www.mistraurbanfutures.org/en

Ariemma, L., & Burnside-Lawry, J. (2016). Transnational resistance networks: New democratic prospects? The Lyon-Turin Railway and no TAV movement. In *Protest, social movements and global democracy since 2011: New perspectives* (Vol. 39, Research in social movements, conflicts and change) (pp. 137–165). Emerald.

Calhoun, C. (1992). The infrastructure of modernity: Indirect social relationships, information technology, and social integration. In H. Haferkamp & N. J. Smelser (Eds.), *Social change and modernity* (pp. 205–236). University of California Press.

Cantoni, F., & Pagnone, F. (2020). Megaprojects. A special eye on sustainability to overcome the 'iron law'. In E. Favari & F. Cantoni (Eds.), *Megaproject management: A multidisciplinary approach to embrace complexity and sustainability* (SpringerBriefs in applied sciences and technology) (pp. 1–13). Springer.

Carse, A. (2014). *Beyond the big ditch: Politics, ecology, and infrastructure at the panama canal*. MIT Press.

Carse, A. (2016). The anthropology of the built environment: What can environmental anthropology learn from infrastructure studies (and vice versa)? Retrieved September 14, 2022, from https://aesengagement.wordpress.com/2016/05/17/the-anthropology-of-the-built-environment-what-can-environmental-anthropology-learn-from-infrastructure-studies-and-vice-versa/

Cohen, R. (2015). Extractive desires: The moral control of female sexuality at Colombia's gold mining frontier. *The Journal of Latin American and Caribbean Anthropology, 19*(2), 260–279. https://doi.org/10.1111/jlca.12098

Coutard, O., Hanley, R. E., & Zimmerman, R. (2005). Network systems revisited: The confounding nature of universal systems. In *Sustaining urban networks: The social diffusion of large technical systems* (pp. 1–12). Routledge.

Deleuze, G. (2011). Postscript on the societies of control. In I. Szeman & T. Kaposy (Eds.), *Cultural theory: An anthology* (pp. 139–142). Wiley.

Dodson, J. (2017). The global infrastructure turn and urban practice. *Urban Policy and Research, 35*(1), 87–92. https://doi.org/10.1080/08111146.2017.1284036

Easterling, K. (2014). *Extrastatecraft: The power of infrastructure space*. Verso Books.

Edwards, P. N. (2003). Infrastructure and modernity: Force, time, and social organization in the history of sociotechnical systems. In T. Misa, P. Brey, & A. Feenberg (Eds.), *Modernity and technology* (pp. 185–225). MIT Press.

Filion, P., & Keil, R. (2017). Contested infrastructures: Tension, inequity and innovation in the global suburb. *Urban Policy and Research, 35*(1), 7–19. https://doi.org/10.1080/08111146.2016.1187122

Flyvbjerg, B. (2009). Survival of the unfittest: Why the worst infrastructure gets built—and what we can do about it. *Oxford Review of Economic Policy, 25*(3), 344–367. https://doi.org/10.1093/oxrep/grp024

Flyvbjerg, B. (2014). What you should know about megaprojects and why: An overview. *Project Management Journal, 45*(2), 6–19. https://doi.org/10.1002/pmj.21409

Flyvbjerg, B. (2017). The never-ending growth of megaprojects. Retrieved November 20, 2022, from https://www.afr.com/companies/infrastructure/the-neverending-growth-of-megaprojects-20170724-gxhfag

Frick, K. T. (2014). Pursuing the technological sublime: How the Bay Bridge became a megaproject. *Access, 44*(2), 22–27.

Furlong, K. (2011). Small technologies, Big change: Rethinking infrastructure through STS and geography. *Progress in Human Geography, 35*(4), 460–482. https://doi.org/10.1177/0309132510380488

Gellert, P. K., & Lynch, B. D. (2003). Mega-projects as displacements. *International Social Science Journal, 55*(175), 15–25. https://doi.org/10.1111/1468-2451.5501002
Hetherington, K., & Campbell, J. M. (2015). Nature, infrastructure, and the state: Rethinking development in Latin America. *The Journal of Latin American and Caribbean Anthropology, 19*(2), 191–194. https://doi.org/10.1111/jlca.12095
Hibbard, K. A., Crutzen, P. J., Lambin, E. F., Liverman, D. M., Mantua, N. J., McNeill, J. R., Messerli, B., & Steffen, W. (2006). Group report: Decadal-scale interactions of humans and the environment. In *Sustainability or collapse? An integrated history and future of people on earth* (pp. 341–375). MIT Press.
Hughes, T. P. (1994). Technological momentum. In M. R. Smith & L. Marx (Eds.), *Does technology drive history? The dilemma of technological determinism*. MIT Press.
Jalbert, K. (2016). Tactics of power and empowerment in knowledge-making infrastructures. Retrieved November 20, 2022, from https://aesengagement.wordpress.com/2016/03/28/tactics-of-power-and-empowerment-in-knowledge-making-infrastructures/
Kaika, M., & Swyngedouw, E. (2000). Fetishizing the modern city: The phantasmagoria of urban technological networks. *International Journal of Urban and Regional Research, 24*(1), 120–138. https://doi.org/10.1111/1468-2427.00239
Kasdogan, D. (2017). *Potentiating algae, modernizing bioeconomies: Algal biofuels, bioenergy economies, and built ecologies in the United States and Turkey*. Dissertation, York University.
Larkin, B. (2013). The politics and poetics of infrastructure. *Annual Review of Anthropology, 43*, 327–343. https://doi.org/10.1146/annurev-anthro-092412-155522
Latour, B. (2007). *Reassembling the social: An introduction to actor-network-theory*. OUP.
Latour, B., & Hermant, E. (1998). *Paris, ville invisible*. La Decouverte.
Matthewman, S., & Byrd, H. (2020). From ‘perfect green dream’ to ‘total failure’: The rise and fall of Christchurch’s district energy scheme. *New Zealand Sociology, 35*(1), 177–202.
Meehan, K. M. (2014). Tool-power: Water infrastructure as wellsprings of state power. *Geoforum, 57*, 215–224. https://doi.org/10.1016/j.geoforum.2013.08.005
Monstadt, J. (2009). Conceptualizing the political ecology of urban infrastructures: Insights from technology and urban studies. *Environment and Planning A: Economy and Space, 41*(8), 1924–1942. https://doi.org/10.1068/a4145
Mukerji, C. (2009). *Impossible engineering: Technology and territoriality on the Canal Du Midi*. Princeton University Press.
Prestel, J. B. (2015). Hierarchies of happiness: Railway infrastructure and suburban subject formation in Berlin and Cairo around 1900. *City, 19*(2–3), 322–331. https://doi.org/10.1080/13604813.2015.1015277
Schindler, S., & Kanai, J. M. (2021). Getting the territory right: Infrastructure-led development and the re-emergence of spatial planning strategies. *Regional Studies, 55*(1), 40–51. https://doi.org/10.1080/00343404.2019.1661984
Shatkin, G. (2022). Mega-urban politics: Analyzing the infrastructure turn through the national state lens. *Environment and Planning A: Economy and Space, 54*(5), 845–866. https://doi.org/10.1177/0308518X221087713
Shove, E. (2015). Infrastructures and practices: Networks beyond the city. In O. Coutard & J. Rutherford (Eds.), *Beyond the networked city* (pp. 260–275). Routledge.
Siemiatycki, M., Enright, T., & Valverde, M. (2020). The gendered production of infrastructure. *Progress in Human Geography, 44*(2), 297–314. https://doi.org/10.1177/0309132519828458
Silvast, A., & Virtanen, M. J. (2019). An assemblage of framings and tamings: Multi-sited analysis of infrastructures as a methodology. *Journal of Cultural Economy, 12*(6), 461–477. https://doi.org/10.1080/17530350.2019.1646156
Star, S. L. (1999). The ethnography of infrastructure. *American Behavioral Scientist, 43*(3), 377–391. https://doi.org/10.1177/00027649921955326
Steele, W., & Legacy, C. (2017). Critical urban infrastructure. *Urban Policy and Research, 35*(1), 1–6. https://doi.org/10.1080/08111146.2017.1283751
Taber, P. (2016). An ‘ecological path’ in Ecuador’s Yasuní National Park: On the reflexivity of oil infrastructure. Retrieved November 20, 2022, from https://aesengagement.wordpress.

com/2016/04/19/an-ecological-path-in-ecuadors-yasuni-national-park-on-the-reflexivity-of--oil-infrastructure/

Trovalla, E., & Trovalla, U. (2015). Infrastructure as a divination tool: Whispers from the grids in a Nigerian City. *City, 19*(2–3), 332–343. https://doi.org/10.1080/13604813.2015.1018061

Urry, J. (2006). Inhabiting the car. *The Sociological Review, 54*(1), 17–31. https://doi.org/10.1111/j.1467-954X.2006.00635.x

Winner, L. (1979). The political philosophy of alternative technology: Historical roots and present prospects. *Technology in Society, 1*(1), 75–86. https://doi.org/10.1016/0160-791X(79)90010-1

Winner, L. (1980). Do artifacts have politics? *Daedalus, 109*(1).

Winner, L. (1983). Technologies as forms of life. In R. S. Cohen & M. W. Wartofsky (Eds.), *Epistemology, methodology, and the social sciences*. Springer.

Winner, L. (1986). *The whale and the reactor: A search for limits in an age of high technology*. University of Chicago Press.

Winner, L. (1993). Social constructivism: Opening the black box and finding it empty. *Science as Culture, 3*(3), 427–452. https://doi.org/10.1080/09505439309526358

Zaloom, C. (2003). Ambiguous numbers: Trading technologies and interpretation in financial markets. *American Ethnologist, 30*(2), 258–272. https://doi.org/10.1525/ae.2003.30.2.258

Chapter 5
The Politics of Type 1 Diabetes Management

Nataliya D. Brantly

Abstract This chapter applies Langdon Winner's insights on the politics of technological artifacts to contemporary biomedical technologies for Type 1 Diabetes (T1D) management. As people with T1D rely on continuous monitoring and treatment technologies to survive, these systems, especially artificial pancreas systems (APS), shape more than clinical outcomes. They participate in the production of knowledge, access, and authority within healthcare. By examining the development and use of APS, the chapter investigates how these technologies can embed and reproduce values, reinforce existing power structures, or generate new forms of inequity. At the same time, innovations emerging from citizen scientists and the broader DIY diabetes community highlight alternative models of technological design, governance, and ownership that challenge dominant assumptions. Drawing on science and technology studies, medical sociology, and bioethics, the chapter reconsiders prevailing narratives about biomedical innovation and interrogates whose needs and perspectives guide these systems. Ultimately, it argues for a more reflexive and inclusive approach to technological development in diabetes care, one that recognizes the political dimensions of biomedical technologies and centers the diverse experiences of those who depend on them.

Keywords Politics of biomedical technology · Type 1 diabetes (T1D) · Artificial pancreas system (APS) · Citizen science · DIY innovation · Power structures

N. D. Brantly (✉)
School of Public and International Affairs, Virginia Polytechnic Institute & State University, Blacksburg, VA, USA
e-mail: nbrantly@vt.edu

© The Author(s), under exclusive license to Springer Nature Switzerland AG 2026

A. Fox, D. Cressman (eds.), *Arrangements of Power: Tracing Langdon Winner's Legacy Within and Beyond the Philosophy of Technology*, Philosophy of Engineering and Technology 50, https://doi.org/10.1007/978-3-032-17371-3_5

5.1 Introduction

One of the greatest challenges facing humanity today is the rapid rise in non-communicable (chronic) diseases, which affect millions of people worldwide. Despite significant technological advances over recent decades, public health statistics reveal a troubling trend: the risk factors for chronic diseases are increasing, leading to increased suffering across populations (WHO, 2023). Technology permeates every aspect of modern life, from education and entertainment to business and healthcare. Within this landscape, human health and wellness have become a core area for technological innovation, which aims to improve quality of life and find potential cures for diseases. The quest for a technological breakthrough that could resolve all human ailments continues, with chronic conditions being a critical focus of this ongoing search.

Biomedical technologies are integral to nearly every aspect of chronic disease care, from diagnosis and monitoring to treatment and management. This chapter explores the role of these technologies in contemporary society and their impact on the human condition. Using the case of type 1 diabetes (T1D) management technologies, I demonstrate the complex relationship between technology, bodies, and socio-political arrangements. T1D is an irreversible chronic condition that necessitates continuous use and reliance on biomedical technologies to maintain life and manage disease. This condition also drives innovation to reduce the risks of lifelong suffering and health complications. The case study illustrates that the adoption of disease management technologies increases the complexity of care and shifts existing power structures intricately entangled with efforts of knowledge production, literacy, transparency, and expertise.

To better understand the political and social implications of biomedical technologies, it is essential to explore the construction and use of these technologies and the power dynamics among the actors involved. This chapter leverages Langdon Winner's scholarly frameworks to contextualize and identify the sociopolitical interactions between technology and health. Winner's influential scholarship continues to impact the study of technology and society, urging scholars to examine the sociopolitical implications embedded within technological artifacts. His ideas on technologically-centered ways of living are particularly relevant in chronic disease management, where technologies enable and sustain life. This chapter extends Winner's insights into diabetes management technologies, challenging technological determinism and critically examining how technologies may embody inherent values, power structures, and ethical considerations that influence their implementation and impact on individuals and communities.

T1D is an autoimmune condition that occurs when one's immune system mistakenly destroys insulin-producing cells in the pancreas. Without insulin, blood glucose (blood sugar) levels rise, and the body cannot function properly without insulin injections. Over 100 years have passed since insulin was first discovered and used to manage T1D. Insulin therapy extended the life expectancy of individuals with T1D from a few weeks post-diagnosis to several decades. However, as T1D became a medically manageable disease, patients began to experience irreversible health complications resulting from prolonged elevated blood glucose levels. As patients increasingly began to use technologies to substitute for pancreatic functions, complications such as nerve, kidney, arterial, cardiac, ocular, and gingival damage became more common. This prompted significant efforts to develop new pharmaceuticals and technologies to prevent or halt health complications and achieve better control over disease. As blood glucose regulation has shifted from a natural biochemical process to a technologically managed one, patients and caregivers must contend with numerous factors that influence blood glucose levels. These include controllable elements, such as daily exercise, food choices, medication intake, sleep, and stress, as well as uncontrollable ones like puberty, menstruation, hormonal changes, and weather conditions.

Today, T1D management necessitates the use of multiple biomedical technologies, such as blood glucose monitors (glucometers, continuous glucose monitors), insulin injection devices (insulin pens, pumps, syringes), and complex systems, including the artificial pancreas (APS, closed loop or hybrid systems). Today's most advanced technology is the artificial pancreas system (APS). However, despite its name, the artificial pancreas is not a single device that imitates the function of a healthy biological organ. It is a system of devices, both on and near the body, that work together and interact with outside systems, such as the Internet, to communicate and transfer users' personal health information. The APS incorporates continuous blood glucose monitoring (CGM) and an insulin pump to automate insulin delivery in response to blood glucose data.

Patients rely on these technologies for the duration of their lives. The increasingly innovative and automated management of T1D promises a better life and freedom from many effects of the disease. However, automation of disease management does not eliminate the risks and burdens of disease; instead, it shifts them and creates new ones (Brantly, 2023). These new risks emerge from reliance on biomedical technologies prone to malfunction and failure. Additional burdens are imposed on patients intimately entangled with and dependent on the corporate supply chains of biomedical technologies to sustain life. For example, despite the complexity of the APS itself, each system component requires a separate prescription from a healthcare professional and approval from a health insurance provider. Each also demands ongoing attention, maintenance, troubleshooting, and proper disposal. Furthermore, each device has a different life and time of activation. Biomedical technologies, though seemingly liberating, impose additional burdens on the individuals they are intended to free from the hardships of disease.

The utility of biomedical technologies should not be underestimated. They have an important role to play in patients' lives. Despite their usefulness, many challenges remain, and many people are still left behind. Among them are people who lack access to the latest biomedical technologies, as well as those who, even today, must ration insulin to stay alive because they cannot afford the needed quantity of it (Fang & Selvin, 2023). Consequently, the patients using advanced biomedical technologies represent a subset of the broader T1D population. Focusing on this group might be considered a limitation of the study but also a strength, as it highlights the current direction of T1D innovation and care.

Biomedical technologies are advertised and celebrated for their breakthrough approach to T1D management. They are elevated on a pedestal as the ultimate solution to T1D despite only being an offset for some attributes of chronic disease. There is a need for a greater understanding of these technologies to see where they fail, who they cater to, who is permitted to engage with biomedical technologies, and at what level. This chapter provides a critical awareness of the broader implications these innovations may have on the lives of individuals with diabetes. It aims to address the following questions. What additional burdens do technologies create? Who is involved in shaping the design and implementation of the APS? How do the implicit political dimensions of APS inadvertently reinforce existing power dynamics, exacerbate healthcare inequities, or introduce new ethical challenges? This chapter challenges prevailing narratives of technological solutionism by questioning the legitimacy of the dominant ideas. It brings a different perspective that advocates for alternative power structures in diabetes management, as exemplified by a growing community of citizen scientists working to make T1D technologies more inclusive and equitable.

This chapter proceeds with a review of how power and authority have changed in managing T1D with the introduction of new and increasingly automated technologies. This section explores how diabetes technologies empower patients, highlighting the emergence of collective movements such as the #WeAreNotWaiting initiative and the development and growing adoption of OpenAPS, an open-source artificial pancreas system. Following this is a discussion of the social construction of diabetes management technologies utilizing Winner's scholarly lens. The Social Construction of Technology (SCOT) as a theoretical approach is particularly helpful for understanding the power dynamics among relevant social groups, the different ways they interpret an artifact's meaning, and their influence on the closure of technological design. This chapter concludes with a discussion of the consequences of technological management of chronic disease. It reveals the unintended consequences of innovation and automation that are often overlooked and poorly understood by the general public. This final section shows the barriers to innovation and demonstrates that biomedical technologies do not inherently guarantee equality, justice, liberty, or freedom. Even with advanced automation, suffering persists, and even those with access to the most sophisticated technologies remain concerned about the consequences, complications, and negative outcomes of living with a chronic condition like T1D. This chapter argues that biomedical technologies, like other technologies examined by Winner, can also embody specific forms of power and authority. In chronic disease management, biomedical technologies, such as the APS, change how power is exercised and how people experience biological citizenship.

5.2 Power and Authority in Type 1 Diabetes Management

Winner's influential piece of scholarship, titled "Do Artifacts Have Politics?", begins with the following thought-provoking sentence: "In controversies about technology and society, there is no idea more provocative that the notion that technological things have political qualities." (Winner, 1980, p. 121) Winner claims that technologies can be evaluated not only for their efficiency, productivity, or effects but also for "the ways in which they can embody specific forms of power and authority." (Winner, 1980, p. 121) Such claims challenge the assumption about technological neutrality and provide a framework for thinking about how technologies influence social and political systems. This section parses the complexity of T1D care before examining the changing power and authority dynamics with the emergence of collective movements involving citizen scientists.

The complexity of entanglements in T1D care is multidirectional. It includes biological, technological, and sociopolitical complexity. Biological complexity explains T1D as a multi-system disease where multiple organs and bodily systems are impacted by insulin deficiency, from endocrine and circulatory to nervous and gastrointestinal systems. Meaning that effective disease management requires a comprehensive approach to monitoring and supporting multiple systems, aiming to prevent complications across various organs and bodily functions. The technological complexity reflects the increased automation that intensifies T1D management. A diverse range of devices and technological systems are used to monitor blood glucose levels, manage insulin delivery, and address different physiological aspects of the disease on an ongoing basis. The sociopolitical complexity includes the shifting responsibility and blame for one's health outcomes. This shift is particularly evident in the transition from doctor-led T1D care, confined to hospitals and clinics, to self-managed care by patients. Blood glucose testing can now be done anytime and anywhere with a glucometer or a CGM without the reliance on nursing staff and laboratories. Likewise, insulin delivery does not require injections by the doctor at the clinic. Instead, the patients can do it at their discretion. T1D technologies have shifted the responsibility for T1D care to the patient alongside the blame for the outcomes of that care.

In the process of technological change in T1D management, the role of regulatory experts has shifted from authoritative to advisory. On the one hand, patients are perceived as empowered to make choices about their health and biomedical technologies within the framework of health politics, outlining what can be prescribed and who is responsible for it (Mol, 2008). On the other hand, patients are perceived as incapable and not very smart, creating a power asymmetry between them and physicians (Williams, 2017). This perception discounts patients as possible experts and innovators. The discounting of patients' expertise can be explained by the fact that not everyone is an innovator, not everyone uses DIY (Do It Yourself) technologies, and newly diagnosed individuals require substantial guidance. However, each person already has intimate knowledge about their body, social circumstances, lifestyle, family medical history, and associated challenges. This knowledge is invaluable for clinical decision-making. Additional expertise in diabetes and biomedical technologies develops with time and experience. A significant outcome of this shift in care and responsibility is patients' growing interest and involvement in T1D technologies, including their modification and design.

Today, a global network of citizen scientists and advocates is involved in diabetes education and the design and sharing of open-source technologies under the #WeAreNotWaiting movement. This movement is making a stance for proactive citizen-led innovation to resolve issues pertaining to diabetes without waiting for commercial entities and regulatory experts to do so. "We [at the Open Artificial Pancreas System project (#OpenAPS)] believe that we can make safe and effective APS technology available more quickly to more people, rather than just waiting for current APS efforts to complete clinical trials and be FDA-approved and commercialized through traditional processes" (OpenAPS, 2024a). Similar movements of people organizing and advocating for their needs based on a health condition have occurred in the past. For example, the movement of people with disabilities in the United States highlighted multiple ways in which architectural designs, common structures, and even computer technologies have made it impossible for many to move and participate freely in society, thus excluding them from public life (Petrick, 2015). The term "biological citizenship" refers to how individuals' rights, identities, and responsibilities are shaped by their biological or health status (Rose, 2009). This concept highlights how access to resources, social recognition, and political rights can be tied to one's health condition or genetic makeup. It allows people to organize based on their biological citizenship to press forward based on common interests to enhance their wellbeing, improve access, and advocate for their rights.

The #WeAreNotWaiting movement emerged in 2013 with an effort to organize and make data and technologies more accessible, intuitive, and meaningful for those with diabetes (Omer, 2016). Members of this group are changing what is possible and what is expected of technological users. Patients' desire for more choice, customization, open connectivity, open standards, and the recognition that "your diabetes may vary" made APS possible (Fernández, 2019). Most notably, members of this group developed OpenAPS before an alternative and current FDA-approved APS solutions were introduced. As of March 26, 2024, there were over 3262 self-reported users of different DIY closed-loop implementations (OpenAPS, 2024b). The number of users continues to grow along with the number of DIY technologies. OpenAPS, as a community effort, is open and free for all on a non-discriminatory basis. It emphasizes safety, access, and effectiveness without profit-seeking motivations.

When considering the two social groups here, the balance of power leans in favor of regulatory experts who support a compliance-oriented approach to care. Such an approach to care is antiquated and reinforces existing power dynamics. Patients are more than passive recipients of medical expertise, as evidenced by T1D self-management and independent daily decision-making. Diabetes technologies increase patients' involvement in self-care, enhance the need for patient education, and improve awareness of patients about their bodies, disease, and lifestyle. Innovation also increased patient's interest in diabetes technologies, their use, inner workings, and design. As a result, many individuals became involved in tinkering with the technology, modifying it, and creating new tools to address their specific concerns. These developments also led patients to question technologies like "black boxes" and health data access and use.

The U.S. Food and Drug Administration (FDA), as the main regulatory body for biomedical technologies in the United States, discourages and frowns upon the use of unregulated T1D biomedical technologies (FDA, 2019). The safety and effectiveness of DIY technologies are the main reasons for the FDA warnings, yet there are currently no accessible pathways for DIY innovators to get FDA approval. The community of DIY innovators, users, and advocates continues to expand, along with the increasing number of unregulated, DIY, and open-source biomedical technologies created, modified, and used by patients. Such growth reflects a level of dissatisfaction with available FDA-approved regulated technologies. It also indicates a discrepancy between the provided healthcare and the type of care desired by those with T1D. FDA-approved APS is a system of technologies that conceal their complexity and make them look easier to manage. Yet, they hide the internal components: conceal coding and software as well as obscure who has access to personal health data and for what purpose. Participants of the #WeAreNotWaiting movement want technological flexibility, transparency, and accessibility.

Ultimately, the choice of T1D technologies becomes a political choice and a statement against (or for) existing power structures. Engaging with the #WeAreNotWaiting movement represents a form of political action and participation expressed through highly creative and diverse methods (Jansky & Langstrup, 2023). Those who rely on DIY technologies choose to take on the responsibility for T1D management with greater transparency, technological flexibility, and associated complexity. While this approach may better meet individual needs, it often faces criticism for being reckless or unsafe and receives limited or no support from regulatory experts. Social and political acceptance of DIY technologies requires change and pathways to not only secure FDA approval as a stamp of safety but also to educate regulatory experts so that people who choose this type of technology can also receive informed support. Having more technological choices and options accessible to more people with T1D can have important consequences for the quality of one's health outcomes, for more democratic ways of innovation, and more holistic approaches to chronic disease management.

Building on Winner's statement that "'technologies' are ways of building order in the world" (Winner, 1980), biomedical technologies are efforts of ordering biological processes (blood glucose monitoring and regulation) that also have the power to reinforce existing political structures (FDA-approved APS) as well as challenge authority and social choice (OpenAPS). As a result, different stakeholders in T1D management have unequal power and unequal awareness about technologies, access, and choice. Dana Lewis, co-founder of the OpenAPS project and a person with T1D, writes: "No matter their choice of tools or technologies, people with diabetes SHOULD be supported in THEIR choices. Not choices limited by healthcare providers, who might only suggest specific tools that they (healthcare providers) have been trained on or are familiar with—but the choices of the patient"

(Lewis, 2022). The inability of DIY innovators to secure FDA approval is a barrier to scaling up and expanding access to innovative DIY technologies to those who need them and might not have the ability to get currently marketed and regulated technologies.

Biomedical technologies like the FDA-approved APS are political in a particular way. They define what is available to patients, what is visible, and what patients are allowed to know. They are black boxes designed for an average user with an imagined level of skill, knowledge, curiosity, and drive. The American Diabetes Association (ADA) is prominent in the creation of the annual "Standards of Care in Diabetes," featuring an editorial board dominated by medical doctors and health professionals (ADA, 2024). Experts regularly define who is involved in conversations about setting diabetes care standards and technological requirements. They determine who can solve problems, innovate, and create new technologies for T1D management by setting standards, norms, and barriers. The group of lay experts reviewed here is challenging current approaches to innovation and established standards of care. The #WeAreNotWaiting movement is creating biomedical technologies that are more open and accessible. Therefore, biomedical technologies for T1D are political arrangements because they both unite and divide people. They make T1D visible and subject to regulation and deregulation. They are changing the dynamics between doctors and patients, between patients and their communities.

5.3 Social Construction of Diabetes Technology

The power and authority in T1D care can be further examined by discussing the social construction of diabetes management technologies. The Social Construction of Technology (SCOT) is a theoretical approach and a key method of inquiry within science and technology studies (STS). It is used to study the technological change in society through the relationship between technology and society. This approach urges scholars to identify relevant social groups and evaluate their roles and their problem with an artifact (object or technology) (Bijker et al., 2012). This approach is crucial because different social groups may interpret the same artifact differently. By analyzing these diverse perspectives, SCOT helps scholars gain a more comprehensive understanding of technology, highlighting limitations and potential issues that might have been previously overlooked. The term "interpretative flexibility" describes how different individuals in various contexts experience and interpret the meaning of an artifact in diverse ways. In the context of T1D management, relevant social groups include regulatory experts and lay experts. Regulatory experts—such as public health professionals, clinicians, and policymakers—base their perspectives on evidence-based science, established norms, and professional standards in medicine, healthcare, and research. In contrast, lay experts—comprising patients, caregivers, and advocates—rely on their diverse lived experiences and intimate knowledge of T1D technologies. These two groups interpret T1D biomedical technologies, such as the APS, differently, as discussed in more detail below.

SCOT, as a theoretical and methodological framework, is not without criticism. Langdon Winner offers a critique of SCOT, as well as its values and challenges. He states that the "social consequences of technological choice," how technologies change communities, the quality of life, people's sense of self, and the overall distribution of power in society, are often left out of view (Winner, 1993, p. 368). He also argues for the inclusion of a greater variety of relevant social actors, including those who might be overlooked due to little or no voice in technological change or adoption. Lastly, Winner argues that SCOT scholars do not take theoretical or practical positions on the technologies, human wellbeing, or issues they study (Winner, 1993). Critiques of this kind are particularly valuable for those who want to "understand the place of technology in human experience" (Winner, 1993, p. 367) and seek relevance to make an impact and influence policy. SCOT and its critique raise important questions, including: Who currently is and should be the authority in technological processes and the adoption of solutions for T1D care? Which relevant social groups are recognized in T1D management? Are some relevant social groups more relevant than others, and why? Who directs and defines the closure of technological design and determines the spectrum of available technologies? The concept of closure indicates that technological systems often reflect the interests of powerful groups, as technologies achieve closure when competing designs or uses are no longer considered. This process can marginalize alternative designs that might better serve other populations. Analyzing relevant social groups concerning closure is crucial, as understanding power distribution provides deeper insights into the dynamics of technology.

Regulatory experts are an actor group typically considered when considering biomedical technologies, their design, prescription as medical devices, and use as data collectors and repositories. Representatives of this group usually support the technological neutrality of biomedical devices and regard technologies with a fundamental conviction that guarantees T1D control, better quality of life, and greater equality with more technology (Brantly, 2023). It is also important to note that diabetes treatment is dominated by powerful pharmaceutical and technology companies (Tattersall, 2009), whose priorities have become increasingly interconnected with the practice of medicine and everyday scientific pursuits (Hunt et al., 2021). The discovery of insulin and technological innovation in diabetes care have inspired unwavering belief and faith in the abilities of scientists and powerful corporations to resolve any human illness with more technological innovation. Historical studies on automation lead us to recognize that there is a need to redirect innovation processes in ways that inspire inclusion and democratic principles (Noble, 2011).

Today's medical understanding of T1D is simplistic and focuses on patients through a lens of biomedical technologies to find an "essential, objective, solid reality" while, in the process, losing the sense of the patient as a person (Feudtner, 2003). T1D care is driven by data derived through biomedical technologies and interpreted by regulatory experts to make care decisions. Medical doctors have "learned to think and act through their technologies," which are crucial for the assignment of disease identification and the legitimization of medical attitudes (Wailoo, 1997, p. 1). T1D is a technologically altered disease. The data-driven medical understanding of T1D prevalent among regulatory experts omits the human realities of living with chronic disease and the challenges that still exist or have been created as an outcome of "transmuted" disease (Dumit, 2012). There is a need to seek a better balance between acknowledging the benefits of biomedical technologies and understanding the incomplete and, at times, harmful consequences of technological solutions.

SCOT posits that technologies are shaped by different social factors and groups, including users and non-users, as well as those who refuse the technology altogether (Oudshoorn & Pinch, 2003). Our society is fixated on innovation and frequently overlooks the "technology-in-use." By including an analysis of the technology in use, one can understand that the "one size fits all" method is ineffective for T1D management or those with T1D due to the variability of bodies, lifestyles, and needs. This is evident from the examination of T1D biological communities, the recognition of individual differences, and the lack of universal preset therapy for T1D. The quantitative ways of seeing disease and its management foster fears, struggles, and the "just-out-of-reach capacity of science to radically reconfigure lives" (Bennett, 2019). Data-driven treatment of T1D alone is insufficient to effectively improve patient's quality of life. Many people with T1D choose to adopt open-source technologies, such as OpenAPS, that might be better suited to meet their needs and provide the necessary transparency and functionality. Some choose to reject innovative technologies altogether because of the data overload, overwhelming feelings and distress due to the constant flow of data, and the burnout from constant alerts, decision-making, and attention that diabetes technologies demand. While most see OpenAPS as a medical device, it is also an artifact that materializes the ambitions of patients, caregivers, and advocates as citizen scientists. It reflects their idea and desires.

Lay experts are also DIY innovators who challenge the status quo in diabetes care. Given that patients are responsible for over 95% of their diabetes care (Funnell & Anderson, 2000), their goals and priorities should be front and center in the discussion of diabetes care, technology use, and efforts aimed at innovation. The biological community of DIY innovators, centered around diabetes issues and solutions, is experience-driven and focused on unmet patient needs as the primary problem. The unmet needs are not only biological but also technological, financial, educational, and social ones. This group views T1D technologies as one of the tools of care with useful features and flaws. It also supports patient empowerment, proactive self-care, and highlights "user agency and individual superiority over technology" (Brantly, 2023). Members of this group have first-hand experience of how the disease, daily activities, and biomedical technologies impact their bodies and wellbeing. Jeremy Greene, a historian of medicine, wrote that our understanding of the disease had moved away from the intimate space of patient and doctor towards abstract areas of policy, guidelines, and markets (Greene, 2007). This process distanced our understanding of disease away from human experiences towards impersonal, technological, quantitative, and data-driven approaches. Looking at users and biological communities helps bring back the human experience and realities of living with a technologically dependent chronic condition.

Historical studies of technological change often demonstrate that efforts of innovation to increase the efficiency and effectiveness of processes often come with disappointing results (Bijker et al., 2012). The reason for this is that technological change is driven by various human motives beyond perceived altruistic efforts to help others, including but not limited to self-enrichment, dominance over others, competition, and cost reduction (Winner, 1980). People are so used to thinking of technologies as neutral that we rarely pause to consider the purposes and driving forces for the design and introductions of technologies in the first place. SCOT offers a way to gather and evaluate different perspectives and views of technology, assess technological change, and start thinking of the consequences of a technological arrangement that exceed intended uses and effects.

5.4 Consequences of Technological Disease Management

If we continue to view biomedical technologies as neutral tools of disease management, focusing solely on their prescribed functionality, "we will be blinded to much that is intellectual and practically crucial." (Winner, 1980, p. 125). This perspective risks overlooking the imbalance in technological design and closure and failing to recognize how biomedical technologies, as political artifacts, exert control and demand discipline. Furthermore, biological citizenship highlights how individuals are increasingly defined by their health conditions and engagement with biomedical technologies, shaping their rights, responsibilities, and social inclusion. Understanding the consequences of technological disease management with reflexive flexibility and consideration of relevant social groups is essential to grasping the politics of disease management, particularly in how it intersects with identity, inequality, and access to care.

The two relevant social groups discussed above represent two distinct yet related approaches to understanding diabetes management technologies. The first group, comprised of regulatory experts, relies on data, established norms, and the quantification of disease. This approach embodies a techno-deterministic perspective on T1D management, focusing primarily on the technological aspects. The second group, consisting of lay experts, acknowledges the internal tensions between technology, the human body, and external factors. This approach emphasizes individuality, the diversity of human needs and biological processes, and the challenges that arise when the complexity of chronic disease intersects with technological complexity and the demands of daily life. Both groups are vital knowledge communities, bringing valuable insights to the broader understanding of diabetes management. Overlooking one of them creates a gap in understanding T1D and its associated technologies. Both are important for innovation and devising ways to approach disease outcomes more effectively. While both approaches are rooted in perceptions of risk, the increasing centrality of technology and automation in the lives of those with T1D has meaningfully changed the power dynamics and the expression of T1D biological citizenship.

In contemporary society, there is a prevalent belief that technological development results from internal processes that, once introduced to the public, follow a predetermined path, molding society, driving societal change, and shaping human behavior, cultural values, and social structures. Technological determinism is the term used to describe this phenomenon. It perceives technology as an autonomous force with human agency considered secondary or even subordinate to technological imperatives. It also means "the human tendency to create the kind of society that invests technologies with enough power to drive society" (Winner, 1997). The drive for innovation also means that fewer individuals dare to challenge conventional ways of technological design. However, the group of citizen scientists driven by the #WeAreNotWaiting movement defy this trend by taking on greater risks and questioning the status quo. The view of the social circumstances of technological development, deployment, and use helps scholars to understand these practices (Winner, 1980). Here, the driving forces of DIY innovations include cost, unmet needs, a lack of access, and a strong will to open the "black box" to provide transparency, visibility, and control over technology to the user.

Winner wrote that a dependency on complex technological systems can also be a source of vulnerability (Winner, 2024a). This vulnerability was particularly evident during the COVID-19 pandemic and the supply chain disruptions that followed. Disruptions of one or multiple components of the systems we rely on undermine the comfort and standard of living many are accustomed to. The idea of dependency as vulnerability can be applied to studying biomedical technologies. T1D technologies are not only networked, interacting with outside technologies and systems but also connected to human bodies. As a result, vulnerabilities associated with technological malfunctions, component failures, or supply chain disruptions threaten not only the quality of patient's lives in terms of comfort and standard of living but also their health and ability to live. A 2019 Dexcom service outage is an example of dependency as a vulnerability. It demonstrates how the reliance on continued functionality of biomedical technology can threaten one's life and safety. Dexcom's "remarkable" revolutionary and "life-changing" technology experienced unexpected server overload that caused the failure to send alerts about dangerous blood glucose fluctuations at night and placed many, especially young children, at risk of death in their sleep (O'Connor, 2019). A technology that promises peace of mind, freedom, and ease of T1D management now makes patients and caregivers alike second-guess technology and plan for technological failure with alternative methods of T1D care, thus increasing the burdens of disease management.

New groundbreaking technologies are often proclaimed to be "the salvation of free society" (Winner, 1980). Similarly, the APS is often described as a freeing, life-changing, revolutionary solution to T1D. While the level of technological complexity and automation depends on accessibility and choice, the increasing demands of care have also increased the burdens associated with T1D self-management. Laura Forlano, a scholar with T1D, wrote: "It is almost ironic that the system that has nearly eliminated the frequent episodes of extremely low blood sugar that woke me in the middle of the night almost a decade ago has enforced another form of sleep interruption and deprivation" (Forlano, 2023). Forlano refers to technological demands for calibration, frequent alarms, and other disruptions. She said: "my body is networked and dependent on a system of technologies that is fragile, vulnerable, and prone to breaking down" (Forlano, 2017). These observations highlight the risks present within the technological component of T1D management and the user's dependency on industrial agendas that introduce interruptions, depravations, vulnerabilities, and additional risks.

The "attitude of trust" (Winner, 2024a) that users develop with their biomedical technologies is supported by corporate discourses. The front page of the Dexcom website alone promises to "simplify diabetes management" with "exceptional confidence," "exceptional accuracy," "unmatched freedom," and "allowing you to live a healthier, more confident life" (Dexcom, 2024a). Throughout its website, commercials and product descriptions show that Dexcom CGM promises "no fingersticks" to draw blood for testing blood glucose levels. Yet, in the small type, it also offers a disclosure that "[f]ingersticks [are] required for diabetes treatment decisions if symptoms or expectations do not match readings" (Dexcom, 2024b). These contradicting claims should undermine trust and any reasonable expectations of absolute reliability, accuracy, and freedom. Biomedical technology users should adopt a healthy dose of skepticism about their technologies. Trust requires knowledge and transparency, both of which are currently lacking. Overreliance on biomedical technologies can be harmful in the case of technological malfunction. It can also come at the detriment of skills and independent thought for alternative methods of care. In the event of system failure, patients might need to revert to manual testing and calculation. Additionally, the promised freedom with increased automation might also require increased literacy and skills related to technological maintenance and troubleshooting, often omitted in corporate discourses.

Technological developments in T1D management changed the requirements and the responsibilities of patients. While patients today have better tools of care and more choices with greater accuracy, biomedical technologies do not relieve patients from the burdens of disease. Today, T1D management is still a chore that requires "perpetual vigilance," "unremitting labor," and relentless attention and effort (Feudtner, 2003). The discovery of insulin and the introduction of complex biomedical technologies, while saving millions from inevitable deaths, did not free those with T1D from careful dietary control and demanding self-discipline (Bliss, 1982). T1D technologies were designed to make life easy: liberating people from decision-making and labor and saving time. However, whose time and labor are they saving? Whose privileged enough to benefit from T1D innovations, and who is left behind? Innovative T1D technologies add complexity and increase the demands of self-care. They do not offer users worry and trouble-free existence.

Just like other technologies used in everyday life, T1D devices have become "information appliances," tools, and instruments for interconnection and communication (Winner, 2000). The APS collects a large amount of data, which is then organized and presented in a standardized report called the Ambulatory Glucose Profile (AGP), which contains calculated statistics and graphs for convenience. Overreliance on data and biomedical technologies in diabetes management is evident. It is partially out of necessity and the demands of T1D. At the most basic level, patients need insulin and insulin injection technologies. Partially, this overreliance results from sociopolitical pressures and efforts of biomedicalization and normalization. Patients are expected to be diligent and to keep disease, their bodies, and the quantified health metrics under control through biomedical technologies. Biomedical technologies are a way to discipline the patient. However, biomedical technologies cannot function and produce data without human action. They require user engagement to fulfill their promise. Their utility fully depends on users. Thus, the user's role should not be ignored but elevated, as technological success depends on the quality and experience of use, creative approaches to technologies, and technological resistance.

As Noble wrote, technological determinism and the denial of the full potential of people, which carries with it experiences, tacit knowledge, creativity, and skills, establish "a severely impoverished realization of the existing possibilities" (Noble, 2011). Yet, the expression of the full potential of people is frowned upon, especially if it goes beyond the norms and expectations established by regulatory experts, especially in relation to human health. Do all stakeholders have an effective voice in shaping biomedical technologies and how they operate and affect others? The answer to this question is no. Lay experts are creating their own ways of salvation, highlighting shortcomings of current technologies and working to escape the "strategies of social control" (Winner, 2024b) by innovating outside the established boundaries and expectations. Technical systems and automation at large, as exemplified by the Artificial Pancreas System, do not guarantee equality, justice, liberty, or freedom. Lay experts are working to change that by pushing for more reflexive and inclusive approaches to technological innovation in diabetes management.

Regulated biomedical technologies as political tools are used to maintain control, categorize individuals, and reinforce a divide between those deemed to have the expertise and those who are expected to accept prescribed regimens and follow directions. These controls are designed to keep authority figures in power and maintain professional boundaries that protect recognized regulatory experts through implicit behavioral norms. Patients are expected to be engaged but not rebellious. They are expected to follow expert recommendations but not proactively modify prescribed regimens or tinker with biomedical technologies. Patients are expected to follow the preestablished standards of care without challenging them. When innovation and policymaking are dominated by powerful businesses, organizations, and affluent individuals, the principles of a democratic society, equality of rights, and equitable access come under threat (Winner, 2015). If the processes of technological change shape the conditions that influence "people's sense of who they are and why they live together" (Winner, 2006), then it becomes imperative to create conditions for inclusive technological change. Involving lay experts in the innovation process and providing accessible pathways for citizen scientists to contribute would empower everyone to think, learn, and innovate. Embracing successful citizen-led innovation efforts would inspire others to imagine and explore new solutions for complex health conditions. A more inclusive technological innovation process would likely be a win-win situation in which the strengths of both lay and expert communities would be combined to foster more tailored patient-centric outcomes.

References

ADA. (2024). Standards of care in diabetes—2024. *Diabetes Care, 47*, S1–S321. https://doi.org/10.2337/dc24-sdis

Bennett, J. A. (2019). *Managing diabetes: The cultural politics of disease*. NYU Press.

Bijker, W. E., Hughes, T. P., & Pinch, T. (2012). *The social construction of technological systems: New directions in the sociology and history of technology* (Anniversary edition). MIT Press.

Bliss, M. (1982). *The discovery of insulin*. University of Chicago Press.

Brantly, N. D. (2023). Automating health: The promises and perils of biomedical technologies for diabetes management.

Dexcom. (2024a). Dexcom continuous glucose monitoring. Retrieved July 31, 2024, from https://www.dexcom.com

Dexcom. (2024b). How Dexcom G7 continuous glucose monitoring works. Accessed July 26, 2024, from https://www.dexcom.com/g7/how-it-works

Dumit, J. (2012). *Drugs for life: how pharmaceutical companies define our health*. Duke University Press.

Fang, M., & Selvin, E. (2023). Cost-related insulin rationing in US adults younger than 65 years with diabetes. *JAMA, 329*, 1700–1702. https://doi.org/10.1001/jama.2023.5747

FDA. (2019). FDA warns against the use of unauthorized devices for diabetes management. *FDA News Release*. Retrieved August 3, 2024, from https://www.fda.gov/news-events/press-announcements/fda-warns-against-use-unauthorized-devices-diabetes-management

Fernández, C. R. (2019). How the DIY Diabetes Community made the artificial pancreas possible. Retrieved July 30, 2024, from https://www.labiotech.eu/in-depth/diy-diabetes-community-artificial-pancreas/

Feudtner, C. (2003). *Bittersweet: Diabetes, insulin, and the transformation of illness*. The University of North Carolina Press.

Forlano, L. (2017). Data rituals in intimate infrastructures: Crip time and the disabled cyborg body as an epistemic site of feminist science. *Catalyst: Feminism, Theory, Technoscience, 3*, 1–28. https://doi.org/10.28968/cftt.v3i2.28843

Forlano, L. (2023). Living intimately with machines: Can AI be disabled? *Interactions, 30*, 24–29.

Funnell, M. M., & Anderson, R. M. (2000). The problem with compliance in diabetes. *JAMA, 284*, 1709–1709. https://doi.org/10.1001/jama.284.13.1709-jms1004-6-1

Greene, J. A. (2007). *Prescribing by numbers: Drugs and the definition of disease*. Johns Hopkins University Press.

Hunt, L. M., Arndt, E. A., Bell, H. S., & Howard, H. A. (2021). Are corporations re-defining illness and health? The diabetes epidemic, goal numbers, and blockbuster drugs. *Journal of Bioethical Inquiry, 18*, 477–497. https://doi.org/10.1007/s11673-021-10119-x

Jansky, B., & Langstrup, H. (2023). Device activism and material participation in healthcare: retracing forms of engagement in the #WeAreNotWaiting movement for open-source closed-loop systems in type 1 diabetes self-care. *BioSocieties, 18*, 498–522. https://doi.org/10.1057/s41292-022-00278-4

Lewis, D. (2022). *We have changed the standards of care for people with diabetes*. DIYPS.org. Retrieved August 3, 2024, from https://diyps.org/tag/wearenotwaiting/

Mol, A. (2008). *The logic of care: Health and the problem of patient choice*. Routledge.

Noble, D. F. (2011). *Forces of production: a social history of industrial automation*. Routledge.
O'Connor, A. (2019). In weekend outage, diabetes monitors fail to send crucial alerts. *The New York Times*. Retrieved July 26, 2024, from https://www.nytimes.com/2019/12/02/well/live/Dexcom-G6-diabetes-monitor-outage.html#
Omer, T. (2016). Empowered citizen' health hackers' who are not waiting. *BMC Medicine, 14*, 118. https://doi.org/10.1186/s12916-016-0670-y
OpenAPS. (2024a). What is OpenAPS? Retrieved July 30, 2024, from https://openaps.org
OpenAPS. (2024b). OpenAPS outcomes. Retrieved August 1, 2024, from https://openaps.org/outcomes/
Oudshoorn, N., & Pinch, T. J. (2003). *How users matter: the co-construction of users and technologies*. MIT Press.
Petrick, E. R. (2015). *Making computers accessible: Disability rights and digital technology* (1st ed.). Johns Hopkins University Press.
Rose, N. (2009). *The politics of life itself: Biomedicine, power, and subjectivity in the twenty-first century*. Princeton University Press.
Tattersall, R. (2009). *Diabetes: The biography*. Oxford University Press.
Wailoo, K. (1997). *Drawing blood: Technology and disease identity in twentieth-century America*. Johns Hopkins University Press.
WHO. (2023). *World health statistics 2023: monitoring health for the SDGs, sustainable development goals*. World Health Organization.
Williams, N. (2017). Fifty years of living with type 1 diabetes. *Clinical Diabetes: A Publication of the American Diabetes Association, 35*, 331–332. https://doi.org/10.2337/cd17-0089
Winner, L. (1980). Do artifacts have politics? *Daedalus, 109*, 121–136.
Winner, L. (1993). Upon opening the black box and finding it empty: Social constructivism and the philosophy of technology. *Science, Technology & Human Values, 18*, 362–378. https://doi.org/10.1177/016224399301800306
Winner, L. (1997). An excerpt from "technological determinism: Alive and kicking?". *Bulletin of Science, Technology and Society, 17*, 49–52.
Winner, L. (2000). The voluntary complexity movement. *Science and Culture, 9*, 103–107. https://doi.org/10.1080/095054300114341
Winner, L. (2006). Who will we be in cyberspace? *The Information Society, 12*, 63–72. https://doi.org/10.1080/019722496129701
Winner, L. (2015). Facing the plague: Economic and political inequality. Retrieved July 25, 2024, from https://www.langdonwinner.com/other-writings/2015/02/facing-plague-economic-and-political.html?rq=Facing%20the%20Plague%3A%20Economic%20and%20Political%20Inequality%20
Winner, L. (2024a). Trust and terror: the vulnerability of complex socio-technical systems. *Science and Culture, 13*, 155–172. https://doi.org/10.1080/0950543042000226594
Winner, L. (2024b). Science policy for true believers. *Conservation Biology, 18*, 866–866. https://doi.org/10.1111/j.1523-1739.2004.01840.x

Chapter 6
Power and Politics in the Design of Autonomous and AI-Enabled Weapons

Nathan Wood

Abstract A weapon is a weapon, and nothing about its general "weaponness" seems to indicate much about its underlying values. However, closer examination of weapons' designs can tell us much. For whom were they designed? Have safeguards been put into place? If so, which ones, and to the protection of whom? Autonomous weapons have existed for decades, and AI-powered and AI-enabled systems are becoming increasingly important, but all of these systems raise ethical and legal concerns. Arguably, many of the challenges might feasibly be addressed through design or institutional solutions, but importantly, *how* systems are designed says a great deal about those designing them and those contemplating their deployment. In this chapter, I argue that if states wish to demonstrate their commitment to human rights and the protection of innocent lives, autonomous and AI-enabled systems in the military must fulfill several design desiderata: predictability, governability, responsibility, and traceability. I then explore concrete ways these desiderata may be fulfilled in practice.

Keywords Autonomous weapon systems · AI · Design ethics · International humanitarian law · Military ethics

N. Wood (✉)
Institute for Air Transportation Systems, Hamburg University of Technology (Technische Universität Hamburg, TUHH), Hamburg, Germany
e-mail: nathan.wood@tuhh.de

Ethics + Emerging Sciences Group, California Polytechnic State University San Luis Obispo, San Luis Obiso, USA

Center for Environmental and Technology Ethics - Prague, Prague, Czechia

© The Author(s), under exclusive license to Springer Nature Switzerland AG 2026

A. Fox, D. Cressman (eds.), *Arrangements of Power: Tracing Langdon Winner's Legacy Within and Beyond the Philosophy of Technology*, Philosophy of Engineering and Technology 50, https://doi.org/10.1007/978-3-032-17371-3_6

6.1 Introduction

A knife is a knife, a bow, a bow, and an ax, an ax. These things may be used in good, bad, or neutral ways. They may be used as tools, and, more importantly, they may be used as weapons, and nothing about the general "weaponness" of a blade, bow, or ax seems to tell us much about the underlying values held by the people who made these things. The same would also appear to hold true for the wide array of modern weapons designed specifically for warfare, from grenades to tanks and missiles to autonomous weapons. Nevertheless, a closer examination of a weapon's design can tell us much about both it and the values of those who made it. Is a weapon made to be usable by anyone, or would it require special training to operate? Are there safety devices or general safeguards built into it? If so, which ones, and to the protection of whom?

The ways we design artifacts speak to what we are designing *for* and what values we care about when we make some tool or weapon. Because of this, if we wish to demonstrate our commitment to certain core values we (claim to) have, we must ensure that our tools and weapons are designed with those values in mind and to foster and strengthen those values in eventual deployments of weapons. This is especially relevant for AI-enabled technologies or autonomous weapon systems (AWS), which can select and engage targets without contemporaneous human oversight.[1] For these, human training and use might even be less relevant than the design of the system itself, and as such, the values of designers and eventual deployers must be accounted for throughout the design process to ensure that the system itself embodies those values.

In this chapter, I explore how power and politics (can) influence the design of emerging technologies in the military domain. I begin by assuming a separation of "power" and "politics" and showing that, at least in well-functioning liberal democratic states, military technologies should not significantly affect the distribution of power in society, and thus, the core concern ought to be the "politics" or values embodied and embedded in technologies (Sect. 6.2). I then discuss the values that (ought to) guide conduct in war (Sect. 6.3), arguing that if states wish to demonstrate their commitment to human rights and the protection of innocent lives, autonomous and AI-enabled systems in the military must fulfill several design desiderata

[1] Throughout this chapter, "autonomous weapon system" (AWS) is understood in line with the definitions of the United States Department of Defense (DoD), the International Committee of the Red Cross (ICRC), and numerous other states and international (non-)governmental bodies, as systems which have autonomy in the "critical functions" necessary for selecting and engaging targets (International Committee of the Red Cross (2014), p. 5.) and which can select and engage targets without contemporaneous human oversight (International Committee of the Red Cross (2021b), p. 1; US Department of Defense (2023), p. 21). For further definitions and taxonomical work, see Altmann and Sauer (2017); Boothby (2016); Boulanin et al. (2020); Caron (2020); Heller (2023); International Committee of the Red Cross (2021a); Taddeo and Blanchard (2022); Williams (2015); Wood (2023a, 2023b), and Pacholska (2024) for careful discussion of subtle differences between prevalent definitions.

(Sect. 6.4). In particular, they must (at the least) be designed such that competent users will understand the risks of using systems in particular discrete environments (predictability), competent users will be able to adequately control systems' activities within reasonably defined geographical and temporal limits (governability), responsibility for the "actions" of systems[2] will be clearly traceable to individuals (responsibility and accountability), and the underlying processes of systems will be sufficiently transparent that designers and troubleshooters can competently address emergent issues as they arise (traceability). Beyond the design desiderata themselves, it is also critical as we move forward that scholars and designers bear vigilance in assuring that the exact design of autonomous and AI-enabled weapons fosters and entrenches these values *at all stages of design, up through deployment* and *into the post-deployment troubleshooting and decommission phases* of a system's life-cycle. In Sect. 6.5, I show how values may be entrenched through the entire life-cycle of autonomous and AI-enabled weapon systems.

6.2 Power and Politics in Military Institutions and Technologies

Power might be understood in terms of violence, force, or domination. However, all of these terms are distinct[3]; systems can exert force without being dominating, there can be violence without power, etc. In what follows, I follow Hannah Arendt's view that "[p]ower corresponds to the human ability not just to act but to act in concert. Power is never the property of an individual; it belongs to a group and remains in existence only so long as the group keeps together".[4] Given this, when an organization *em*powered by individuals loses its trust (and backing), it loses its power (though it may still have the capacity to mete out violence).[5] On the other hand, politics concerns how power and authority are arranged and for what those arrangements are used.[6] When we consider "an individual's politics", it is also clear that though the arrangement of power and authority can matter to one's politics, the core point generally concerns *what* one values. Thus, for the individual, whether one favors more broadly progressive or conservative policies represents a more potent marker of one's politics than what form of government one finds most appropriate, especially given that in most states and societies, the form of government is largely

[2] See Wood (2025a) for critical discussion of the language of machines "acting" or "deciding".

[3] Though cf. Winner (1978), p. 139, where power and force are explicitly conflated: "Power is ultimately the power of nature itself, released by the inquiries of science and made available by the inventive, organizing capacity of technics".

[4] Arendt (1970), p. 44. See pp. 35–56 for extensive discussion of all these terms with reference to historical interpretations. For thorough examination of "domination" from the republican tradition, see, e.g., Pettit (1997); Skinner (2002, 2008).

[5] *Ibid.*

[6] Winner (1980), p. 123.

fixed. Political questions are centrally questions of value. Thus, asking, "What are your politics?" is akin to asking, "What matters to you?" or "What do you value and want promoted?". Going forward, I will take this understanding of "politics-as-values" for granted.

Regarding military institutions, there are three broad categories of persons over whom militaries might exercise power: their own civilian population, foreign civilian populations, and foreign militaries. In well-functioning societies, especially liberal democratic ones, where civilian authorities control military institutions, militaries will not exercise or hold power over civilian populations. On the contrary, civilian populations empower militaries to act on their behalf, and that power can be recalled should militaries overstep their bounds. Militaries also ought not to be exercising power over foreign civilian populations in general, as it is not the job of militaries to control the civilians of another state.[7] Moreover, though foreign civilians will not have power over another state's military, if that military is exercising power over them, then this arguably constitutes a breach of the international order in need of remedying. The one group where militaries do exercise clear power is over adversarial military forces, but this is not a problem; it is the job of militaries to create imbalances in power and to seek, to the greatest extent possible, to exercise power over adversary forces to compel them to surrender or desert. One may object to that base concern of the military (i.e., accept pacifism), but that is a more significant commitment, the discussion of which falls outside the scope of this chapter. In any event, in well-functioning societies, a military and its equipment should not be altering the balance of power in a troubling way, so the primary question is whether the military and its equipment comport with the society's politics or with their values.

6.3 Values in War

Sadly, warfare has been an enduring aspect of human interaction for millennia, leading to the emergence of basic norms and principles of war across disparate peoples and cultures.[8] These began as moral or religious rules but have since become codified into international humanitarian law (IHL), the legal regime governing conduct in war. In IHL and the ethics of war, four principles form the pillars of our values in war: distinction, necessity, proportionality, and dignity/humanity.[9]

[7] Occupation during or after conflict raises some issues, as in these cases militaries will take on temporary law enforcement roles. However, in that event, the military should not be using kinetic combat systems in many, if any instances, as the military is then serving as a police force and not a military (vis-à-vis civilians).

[8] For broad historical overviews of varying traditions and views, see e.g., Bellamy (2006); Brunstetter and O'Driscoll (2018); Cox (2018); Kalmanovitz (2018); Kelsay (2007); Reichberg (2018); Schwartz (2018); Sorabji and Rodin (2006).

[9] These principles can all be found in the Hague and Geneva Conventions (as well as the Additional Protocols) which enjoy almost universal support and set the basis for customary IHL which is

Distinction is perhaps the most elementary principle of warfare, maintaining (very roughly) that one is required to direct all violent action against legitimate military targets and spare, to the extent possible, civilians, soldiers who are *hors de combat*, and other protected persons.[10] Distinction encompasses the basic value that harm ought not be dealt to those who are innocent, defenseless, or not participants in conflict, and the principle is thus about *distinguishing* between legitimate and illegitimate targets. In practice, this value is embodied through training and doctrine guiding combatants to only fire when they are confident their targets are legitimate ones, and they can direct their fire well. The contemporary understanding of distinction also rules out the use of weapons or tactics that preclude one from adequately distinguishing between legitimate targets and protected persons or objects.[11] Regarding weapons design, distinction requires that weapons not be of a nature to preclude their being used in a discriminate manner.[12] There is much more that can be said regarding distinction,[13] but the above rough characterizations will suffice for the purposes of this chapter.

Now, supposing one has adequately distinguished between legitimate military targets and those who ought to be spared the harms of war, directing one's attacks only toward the former, it is at this point that necessity rears its head. At a basic level, necessity demands that one only imposes those harms necessary to achieve one's aims.[14] More practically, necessity enjoins one to minimize harm to the extent possible when conducting military operations, as any harm beyond that minimal required amount is not strictly *necessary*. It is difficult to determine with high confidence what level of harm is both minimal yet still sufficient to affect one's aims, but as a general rule, necessity leads one to explore less harmful means and to scale up attacks as needed rather than starting with massively harmful approaches. With regards to design, necessity forbids the development and use of weapons that would

---

binding for all states.

[10] For distinction in IHL, see Geneva Protocol I Additional to the Geneva Conventions (AP I), Arts. 41, 48, 51, and 57, Hague Convention IV, esp. Art. 23, Art. 3 of Geneva Conventions I–IV, and the broader discussion of Rule 47 of the ICRC Customary IHL Database (https://ihl-databases.icrc.org/en/customary-ihl/v1/rule47). See also Sandoz et al. (1987), p. 598, paragraph 1863; Solis (2016), p. 309; US Army Judge Advocate General (2015a), pp. 137–138; US Army Judge Advocate General (2015b), pp. 17–19.

[11] See AP I, Art. 51.4–5.

[12] For example, biological weapons represent a type of weapon which likely cannot be directed in any reasonably adequate fashion, and which are thereby forbidden on grounds of distinction (in addition to other grounds which may forbid the use of such weapons).

[13] See, e.g., Blanchard and Taddeo (2022); Haines (2014); Pacholska (2023); Wood (2025b).

[14] For extensive discussion of the intricacies of necessity in both the ethics and law of war, see respectively the 1868 St. Petersburg Declaration; AP I, Arts. 35, 57; Daoust (2001); Dill and Shue (2012); Gardam (2004); Haque (2018); International Committee of the Red Cross (1997); Ohlin and May (2016); Onishi (2018); US Army (1956), paragraph 2; US Army Judge Advocate General (2015b), p. 14; and Goodman (2013); Lazar (2012, 2013); Mayer (2013); Ohlin (2012); Statman (2011).

cause unnecessary suffering or superfluous injury,[15] arguing instead for weapons and methods of war that are forceful enough to allow one to achieve one's aims while being as limited as possible in their harmful capacity.

Related to necessity is the principle of proportionality, which maintains that the harms imposed for the sake of achieving some objective ought not be disproportionate to the value of the objective itself.[16] Within IHL, this is understood as forbidding the imposition of harm to civilians "which would be excessive in relation to the concrete and direct military advantage anticipated" by an attack.[17] Importantly, proportionality and necessity can come apart in practice, as it is possible for some objective to only be achievable through the use of some method or tactic (i.e., necessary harm) but for that method or tactic to be too harmful to be permissible (i.e., disproportionate harm). Considering both necessity and proportionality at a general level, proportionality may be understood as a basic comparison (i.e., "Is the objective worth *this much* harm?"), while necessity can be seen as an optimizing principle (i.e., "Can we achieve the objective *with less* harm?"). As proportionality is primarily concerned with choices about attacks and the methods and tactics used, it has less to say about concrete design decisions for weapons. That said, proportionality serves to inform the broader set of values with which we design weapon systems.

The final core value in war is the notion of dignity or humanity. As dignity/humanity is a concept that can be hard to define (and even harder to get a broad set of individuals to agree on),[18] it serves to ground fewer concrete rules or principles of conduct. However, it allows us to examine what types of weapons or conduct we view to be fundamentally disrespectful to us as humans and can serve as a guiding light in discussions of how we ought to comport ourselves. Thus, despite a certain lack of concreteness, dignity or humanity can act as a foundational value in checking our design decisions, forcing us to ask the broad (and deeply philosophical) question of whether some weapon is fundamentally inhumane or in violation of a person's dignity (however these may be understood).[19]

Many other rules and norms guide conduct in war, but these represent some of the most foundational principles and are the most widely accepted. Yet in addition to these values, a further value cannot be ignored, namely effectiveness. For all of

---

[15] The notion of "unnecessary suffering and superfluous injury" forms the core of the legal conception of necessity can be found in the very foundations of IHL. See, e.g., the 1868 St. Petersburg Declaration, as well as AP I, Art. 35, which states necessity under the heading of a "Basic Rule" of war.

[16] See, e.g., Hurka (2005, 2008); McMahan (2013, 2015, 2018); Steinhoff (2017); Tomlin (2019, 2020); Uniacke (2011).

[17] AP I, Art. 51.5.b.

[18] See, for example, the variety of definitions within the philosophical literature: e.g., Griffin-Heslin (2005); Killmister (2010); Michael (2014); Nordenfelt (2009); Rosen (2012).

[19] In the law, methods or means of combat which are deemed mala in se, or evil in themselves, are generally taken to be prohibited on grounds of dignity (among other concerns). Such prohibited methods and means traditionally taken to meet this definition would be, among others, rape as an instrument of war (AP I, Arts. 75–76), torture (AP I, Art. 75), or the use of undetectable fragments (Protocol I of the 1980 *Convention on Prohibitions or Restrictions on the Use of Certain Conventional Weapons Which May be Deemed to be Excessively Injurious or to Have Indiscriminate Effects*). See also Dige (2017); Rosert and Sauer (2019).

the humanitarian values espoused in the ethics and laws of war, the simple fact is that war is still war. It is still an enterprise aimed at physically besting an adversary, which usually entails the maiming and killing of one's enemies on a massive scale. The preceding values aim to limit the destructiveness of war, but they exist in constant tension with the need for an effective and efficient military. And this entails having armaments that can be highly destructive, that can be extremely lethal, and that allow one to execute violent action quickly and without reservation when needed. In designing weapons of war *with* and *for* values, this tension must be constantly borne in mind, as weapons must balance both military necessity and the humanitarian concerns intended to minimize the harm of war. The following section will explore how this may be done in practice: how designers may seek to entrench ethical and legal values while still allowing for systems which are sufficiently effective to provide states with defensive capabilities.

6.4 Designing Weapons *with* and *for* Values

As a first point, it is worth stating clearly that no matter how weapons are designed, humans deploying them bear overriding responsibility for using them well, per the ethics and law of war. Combatants must also employ their arms *effectively*, which unfortunately usually implies the grave (and often lethal) wounding of other humans, but choices may be made regarding how, where, when, and with what precautions weapons are used to minimize harm to both the enemy and any non-combatants present. Thus, humans employing weapons ultimately have the "final say" regarding what values are respected and furthered during wartime. How weapons are designed can lead one to better or worse outcomes and to better or worse general attitudes regarding norms of war. More fundamentally, the design of artifacts speaks to and demonstrates our higher-level politics and our foundational ethical, legal, and social values.[20]

If autonomous and AI-enabled weapons are designed as "point-and-shoot" devices, where combatants need only deploy them and await their enemies' destruction, then this would likely have many pernicious effects. From making warfare potentially far more lethal to gravely increasing the risks of mistaken targeting of protected persons, the more easily an AWS or AI-enabled system may be deployed, the more likely it is that combatants will make use of them without due care and respect for all those put at risk through the use of such systems. However, if these

---

[20] Winner (1980, 1986). Winner's work also serves as the foundation for the "Value Sensitive Design" (VSD) approach to ethical innovation. For exploration of VSD in general and with respect to autonomous weapons, see, e.g., Dexe et al. (2020); Friedman (1996); Friedman et al. (2002, 2013, 2017); Friedman and Hendry (2019); Reijers and Gordijn (2019); Saket (2021); Umbrello and van de Poel (2021); van den Hoven (2013); van den Hoven and Manders-Huits (2020) and Boshuijzen-van Burken (2023); Sancar (2024); Umbrello (2022); Verdiesen and Dignum (2023); Wyatt and Galliott (2021).

systems are designed such that combatants must meaningfully engage with each deployment decision, accounting for various values we hold dear in war, risks may be mitigated without any necessary loss of military capability. Forcing such meaningful engagement also does not imply that combatants need be "in-the-loop" or "on-the-loop", that is, directly contributing to an autonomous system's decision-making or contemporaneously overseeing that decision-making.[21] Instead, systems need only be designed so that combatants are presented with information and a user interface that demands they account for the many factors that impact the responsible, ethical, legal, and effective use of autonomous and AI-enabled weapons. This means that these systems are, at base, predictable enough for combatants to make reasonable assessments regarding the expected (and expectable) outcomes that will follow a deployment decision. They must also be designed to be adequately controllable or governable by combatants, in that they are not to be "taken off the leash" broadly or strongly. Autonomous weapons must be tightly tied to discrete operations with discrete aims in discrete environments. Such governability also naturally links the deployments of AWS and AI-enabled systems, and their outcomes, to combatants, providing a post facto check on values (more on this below). And finally, systems may be designed to make it easier or more likely that mistakes and problems may be identified and addressed, making it so combatants are better positioned not just to use such systems in the first instance responsibly but also so that they may responsibly respond to failures of such systems or failures of deployment.

Approaches to designing weapons *with* and *for* values are not new, and indeed, such approaches have already been applied in discussions of autonomous and AI-enabled weapon system design.[22] However, many of these discussions focus on the "nuts and bolts" of ethical design, on eliciting values and creating design programs that keep these values centered. Often overlooked in such discussions, though, is the importance of the realities of warfare, of how operational and contextual elements impinge on what values we can seek to further in war, and, more importantly for our purposes here, how the realities of war may undermine certain design choices. To be more precise, abstract discussions of "designing for values" may provide one with general ideas on how design may further some set of values. However, when weapons are deployed to a theater of conflict, the realities of that particular theater may alter the realized outcomes in ways designers may not have foreseen or even been able to foresee well (perhaps because designers are, well, designers, and not military professionals with a relevant understanding of operational needs and realities). By focusing not just on values and design but more concretely on how wars are fought, what challenges combatants face, and what contextual elements may alter the potential outcome space when deploying a particular asset, we may seek to gain a better understanding of how we can robustly and reliably design *with* and *for* values. In particular, to foster, entrench, and further

[21] Crootof et al. (2023); Heins (2018); Ivanov (2023); Wood (2025a).

[22] E.g., Boshuijzen-van Burken (2023); Sancar (2024); Umbrello (2022); Verdiesen and Dignum (2023); Wyatt and Galliott (2021).

values in war, we must design autonomous and AI-enabled systems such that they are, at a minimum, sufficiently predictable and governable, they track clear lines of responsibility for their use, and their internal structures and processes are adequately traceable to designers and troubleshooters. It is to these design desiderata that we now turn.

6.4.1 Predictability

If our (use of) weapon systems are to respect, further, and entrench the principles of distinction, necessity, proportionality, and dignity in war, all while affording combatants the greatest combat capabilities, a minimum design criterion is that systems be sufficiently *predictable*. Predictability represents an essential requirement, as it is only with predictable weapon platforms that combatants can even begin to assess what will expectably follow from deployments. Moreover, having such predictions is a precondition for determining whether only legitimate targets will be harmed through the use of a weapon (distinction), whether objectives could be achieved by using less harmful means or methods (necessity), whether expected harms are not disproportionate to the concrete and direct military advantage anticipated (proportionality), whether a weapon's use would expectably violate the dignity or humanity of some individuals during deployment (dignity), and finally, whether the weapon would provide military advantage (effectiveness).

With regards to design, this means that autonomous and AI-enabled systems must be built such that, given discrete inputs or stimuli, systems will predictably and reliably deliver specific outputs. More concretely, systems should "[f]unction as anticipated in realistic operational environments against adaptive adversaries taking realistic and practicable countermeasures" and the systems' "capabilities will have explicit, well-defined uses, and the safety, security, and effectiveness of such capabilities will be subject to testing and assurance within those defined uses across their entire life cycles".[23] This might be argued to entail that systems which exhibit any *un*predictability undermine our values in war. Building on this, systems that are constructed using opaque or "black box" architectures may also be argued to present a problem, as such systems possess an inherent capacity for unpredictable action, given that users (and even designers) do not and cannot fully see how the system goes from inputs to outputs. Thus, it may be argued that the design desideratum of predictability, if it is to bolster our values in war, rules out the development and deployment of systems which are opaque or which retain an inherent capacity for unpredictable behavior. However, this would be too quick.

First, having an "inherent capacity for unpredictable behavior" is not the same as "being unpredictable"; it is often the case that we have systems that *could* act in unpredictable ways but that, *in practice and actuality,* comport with our

[23] US Department of Defense (2023), pp. 4, 6.

expectations at virtually every juncture. And considering that predictability is also often tied to opacity in discussions critical of AWS and military AI, it is also worth considering how opacity does *not* lead to predictability concerns. Perhaps the best illustration of this is the fact that we deploy both human and animal combatants in warfare and can trust, rely on, and reliably predict how they will act in many (if not most) environments, despite the fact that humans and animals are fundamentally opaque and capable of acting in unpredictable ways.[24] But, as should be evident, that a person or animal *could* act unpredictably and in unwanted ways by no means implies they *will* do so. Determining whether some entity or thing is predictable is thus less about opacity and instead focuses on reliability assessments in testing environments.[25]

Consideration of such assessments illuminates a further aspect of predictability, namely that it is both user- and context-dependent. At a basic level, we can determine some generic "level of predictability" for a system, but that generic assessment will, in many, if not most cases, be unhelpful or potentially even inaccurate. This is because a system's predictability or reliability can have more to do with where, when, how, and by whom it is being used than because of anything about the system itself. And one system which is more predictable than another in a generalized sense may be less predictable in particular deployments, because the second system is perhaps designed to handle challenges posed by the particular deployment environment in question. Beyond this, systems may be highly predictable when used by competent users but highly unstable and unpredictable when used by novices. Thus, predictability is more about the use of a system than about a system itself.[26]

That said, there are design choices that may further or strengthen the predictability of autonomous and AI-enabled weapons. Given the above points, that context and user characteristics may often matter more than systems themselves, we can promote predictability (and thereby the deeper values requiring predictability) by designing systems that demand a high degree of competence in order to be used at all. Thus, rather than having "point-and-shoot" devices, we may design AWS such that only those given a high degree of training with the particular system in question can use it at all. This would reduce the risks of novice users recklessly or negligently deploying such systems and would moreover set a higher standard for human-machine teaming, a fact which in and of itself can help foster and entrench

[24] See Wood (forthcoming).

[25] This leads to a slight quirk in the "design desiderata" of predictability, namely that predictability is not so much a design choice as it is a characteristic determined through testing and evaluation (T&E) and verification and validation (V&V). Thus, we do not "design predictable systems" so much as we "design systems and then determine whether they are sufficiently" predictable. Yet even though this speaks against the idea of designing *for* predictability, if we are concerned with the entire life-cycle of autonomous and AI-enabled weapons—as we should be—then predictability is something which is examined in testing, deployment, post-deployment troubleshooting, and eventual repairs/reprogramming of systems, and to that extent can be accounted for within the context of a comprehensive design approach.

[26] For elaboration on this argument, see, e.g., Wood (2024b).

responsible and effective use in a variety of significant ways.[27] Thus, to promote predictability, there may be good grounds to, somewhat unintuitively, design systems that are harder to use, require more technical experience, and provide users with less straightforward guidance, all to ensure that users have a deep understanding of the systems they will be using. At any rate, design decisions aimed at promoting predictability must look not at laboratory predictability assessments a system might be given but instead focus on how real combatants in real warfighting scenarios will be deploying systems and what design choices would aid those real individuals in reliably making use of the systems they are given.

6.4.2 Governability

Related to the aspects of predictability just discussed, for combatants to effectively use autonomous and AI-enabled systems while respecting and promoting humanitarian values in war (distinction, necessity, proportionality, and dignity), these systems must be sufficiently *governable*. Fundamentally, governability means that systems must be subject to meaningful human control.[28] Meaningful human control may be narrowly understood as requiring that humans be in- or on-the-loop throughout AWS deployments, or it may more broadly be understood as requiring deployments be appropriately tied to human decision-making and agency to ensure that the AWS is serving as an extension of the human's will.[29] More concretely, the narrow understanding would hold, on the one hand, that designers ought to "design and engineer AI capabilities to fulfill their intended functions while possessing the ability to detect and avoid unintended consequences, *and the ability to disengage or deactivate* deployed systems that demonstrate unintended behavior".[30] The broad understanding, on the other hand, might only require that deployments be temporally and geographically constrained such that an operator/deployer exercises judgment and governance to ensure that all and only intended targets will be engaged due to the targeting and deployment parameters set. This looser understanding allows for some degree of risk to be taken by combatants deploying AWS and AI-enabled systems but limits that risk by demanding that combatants tie deployments to highly discrete times and places.

Whether one follows a narrower or broader understanding of "meaningful human control", governability requires, at the very least, that AWS "[c]omplete

[27] For recent discussions of human-machine teaming in the military, see Cappuccio et al. (2021); Johnson (2023a, 2023b, 2024); Macak (2005); Neads et al. (2021); Roff and Danks (2018); Scharre (2016); Wood (forthcoming).

[28] For discussion of meaningful human control in debates on autonomous and AI-enabled systems in the military, see, e.g., Bode and Watts (2021); Ekelhof (2019); Human Rights Watch (2016a); Roff (2024); Santoni de Sio and van den Hoven (2018); Umbrello (2021a, 2021b); Wood (2025a).

[29] Wood (2025a).

[30] US Department of Defense (2023), p. 6, emphasis added.

engagements within a timeframe and geographic area, as well as other relevant environmental and operational constraints, consistent with commander and operator intentions. If unable to do so, the systems will terminate the engagement or obtain additional operator input before continuing the engagement".[31]

Regarding the design of these systems, this leads to natural decisions such as requiring that, for an autonomous system to be deployable, it must have a clearly demarcated area of operations selected. We may further require that strict mission timeframes be incorporated at the design stage, making it impossible for systems to be put on open-ended or extended missions. This would tie AWS deployments to narrow operational windows, making them assets that, by necessity, would be intimately tied to commanders' intentions, bringing several benefits.

First, by requiring strict temporal and geographic deployment parameters to be set, commanders would be forced to carefully consider exactly what timeframe and area of deployment is necessary. These parameters alone would help anchor the humanitarian values guiding conduct in war by forcing combatants to think about missions in a limiting and minimizing way. Additionally, in the case of narrower views of meaningful human control, if one requires that systems be capable of manual override or "obtain additional operator input before continuing", this further entrenches human judgment, promoting governability and the deeper values of war. Finally, by requiring tight temporal and geographic deployment windows to be set, the exact outcomes relating to deployment decisions (and their attendant parameters) could be more clearly discerned, making attributing blame or guilt (in the event of mistakes) easier and more likely. Which brings us to the next design desiderata, responsibility and accountability.

6.4.3 Responsibility and Accountability

One long-discussed objection to autonomous and AI-enabled weapons is that they will create so-called "responsibility gaps", where their use makes it impossible to determine who should be held responsible when mistakes are made, leading to outcomes potentially constitutive of war crimes.[32] There is much debate concerning whether such gaps exist and, if so, whether they can be mitigated through technical and institutional solutions,[33] but regardless of these debates, it is clear that in order for our values in war to be respected and furthered, we must be able to *identify individuals responsible* for acts of war and *hold them accountable* when things go wrong. If we cannot, then no value may be enforced, and general norms are likely

[31] *Ibid.*, p. 4.

[32] For discussions from both ethical and legal perspectives, see respectively Asaro (2006); Human Rights Watch (2015); Matthias (2004); Pagallo (2011); Sharkey (2007); Sparrow (2007); Stop Killer Robots (2018) and Chengeta (2016); Crootof (2016); Wagner (2014). For responses, see, e.g., Hindriks and Veluwenkamp (2023); Königs (2022); Oimann (2023a, 2023b); Wood (2023b).

[33] See preceding citations.

to weaken. Without taking any stand on foundational philosophical and legal debates surrounding responsibility gaps, we can implement several concrete design choices to ensure that "personnel will exercise appropriate levels of judgment and care, *while remaining responsible for the development, deployment, and use of AI capabilities*".[34]

To begin, it is worth highlighting that all design choices that promote predictability and governability will also further responsibility and accountability by virtue of making it clear what will follow from deployments; when combatants have reliable knowledge about what a system will predictably do in a given environment, must set clear targeting parameters, and delimit the time and space of operations, it becomes difficult to maintain that potential mistakes could not reasonably be foreseen. These elements tie the outcomes of deployments to combatants' deployment choices in integral ways, anchoring responsibility and making it much easier to hold combatants accountable should foreseeable or preventable mistakes be made. Responsibility may be further designed for in (at least) three ways.

First, autonomous and AI-enabled weapons are almost certain to be designed with cameras and other sensors capable of processing troves of visual and other data. This is simply because such systems, if they are to be effective and useful, will have to possess (multi-modal) means of identifying potential targets. Taking this for granted, systems may be designed to not only *process* this data but also *record or transmit it*, creating a log of all that systems see, hear, or otherwise sense. Logs could then be used to assess what went wrong if mistakes were made, allowing for the assessment of who or what caused the failure. Knowledge that a running log is being held would also incentivize combatants to act with more judgment and discernment and, moreover, act as a check against potentially bad actors who might otherwise be apt to misuse systems.[35]

Related to this and building on the points regarding governability, systems could also be designed to have "black boxes" that hold the data collected during a mission and that further record the temporal and geographic limitations set by combatants deploying the system, as well as any targeting or mission parameters combatants might provide. Such "black box" data recorders would need to be designed to preserve data in an unalterable or non-tamperable state, creating a permanent log of

---

[34] US Department of Defense (2023), p. 6, emphasis added. The document's preceding discussion of policy recommendations provides further concrete points on how predictability, governability, and responsibility may be furthered through design choices, such as that "the human-machine interface for autonomous and semi-autonomous weapon systems will:

(a) Be readily understandable to trained operators, such as by clearly indicating what actions operators need to perform and which actions the system will perform.
(b) Provide transparent feedback on system status.
(c) Provide clear procedures for trained operators to activate and deactivate system functions." (US Department of Defense (2023), p. 4)

All of these design choices would serve to further anchor responsibility in those combatants who are making use of the system.

[35] Müller (2016).

exactly what precautions, parameters, and priorities were given to the autonomous or AI-enabled system. This logging would also create clear lines of accountability if mistakes were made, entrenching responsibility, making accountability more likely, and thereby promoting and entrenching the humanitarian values of war.

Finally, responsibility may also be furthered by requiring that AWS and military AI only be usable within strongly established and competent human-machine teams. The teaming of humans and autonomous or AI-enabled systems is primarily an institutional affair, calling more on trainers and military educators than on designers, but even so, there are design decisions that could bolster this effort; autonomous and AI-enabled systems could be designed such that only certain individuals possessing sufficient training have the knowledge needed to boot up and deploy systems. Systems could also be specifically keyed to those combatants tasked with handling the systems, requiring either password or biometric verification that those permitted to use the system are indeed the ones at the controls. By linking specific combat platforms to specific users, only those sanctioned to deploy them could, and it would be assured that handlers of the systems would be logged upon startup, creating clear lines of accountability.

All in all, these design choices would create higher chances of responsible use of AWS and military AI, making it difficult for bad actors to deploy these systems surreptitiously. If responsibility is promoted through the direct connection of specific combatants with specific weapon platforms (through rigorous verification methods), this could even go so far as to make it (nearly) impossible for systems to be used without a responsible party, should mistakes occur. On top of this, by having large amounts of data tracking what might go wrong during deployment, good actors who just so happen to have the misfortune of deploying a defective system would also have better chances of avoiding unfair attributions of accountability. After all, no matter how reliable a platform is in general, there will be some number of units that are simply defective, but logs detailing how, where, when, and under what precautions systems are deployed can serve to exonerate combatants who have the misfortune of being paired with defective systems. Finally, if some number of defective units are circulating within a military, records of each system's use would provide clear trails of information to be used in troubleshooting, repairing, and reprogramming of possibly defective systems. These records would further promote our ability to improve systems regarding their accordance with the norms and values of war.

6.4.4 Traceability

The above design desiderata provide numerous ways to foster selected values in both the design and deployment phases of autonomous and AI-enabled systems, with the discussion of responsibility and accountability also indicating some ways in which values may be entrenched post-deployment. *Traceability* complements these desiderata by focusing primarily on the post-deployment side of these technologies, showcasing how we may best design systems that allow for running improvements and alterations to be made and allow for changing values in society and in war. The core of traceability is the conviction that designers and troubleshooters must have the ability to competently address defective systems or emergent problems as they arise, which arguably requires that "AI capabilities will be developed and deployed such that relevant personnel possess an appropriate understanding of the technology, development processes, and operational methods applicable to AI capabilities, including with *transparent and auditable* methodologies, data sources, and design procedures and documentation".[36]

For simpler autonomous systems, traceability can be a straightforward and perhaps trivial matter. This is because many current AWS are point defense systems, loitering munitions, anti-materiel weapons, or active protection systems for mobile assets. These have little to no AI and often function based on simple(r) algorithmic or even conditional reasoning. However, as AWS become imbued with ever more complex and opaque AI systems, they will become less transparent, and their reasoning will become less traceable. To address this, there is a growing body of research on so-called "explainable AI" (XAI).[37] By creating AI systems that supplement their outputs with clear explanations users can engage with, these systems are taken to be more accessible for both users and troubleshooters. However, some argue that explanations may not provide the boon they purport because they can be unhelpful, unwieldy, or even because they may undermine the perceived need for strong human-machine teams,[38] thus detracting from a central need for responsible and effective deployment of autonomous and AI-enabled systems. Nevertheless, even if we do not maintain that AWS and military AI need to be fully transparent or have their reasoning be fully traceable, even incomplete traceability will further values in war.

The more accessible system processes are to users, the more that users will be able to reliably gauge how systems will behave in various environments. This furthers the effective predictability of systems, allowing combatants to better use them. Furthermore, more transparent and traceable systems afford less opportunity for combatants to claim that behaviors were unforeseeable, limiting the possibility for inappropriate invocations of ignorance for the sake of avoiding responsibility attribution when mistakes are made. The main gain from traceability is not to be found on the battlefield, though, but rather in the repair bay.

By designing systems that are as transparent and traceable as possible, when mistakes are made, or systems perform at odds with what users would expect (and perhaps wish), it may not be necessary for users or deployers to directly see what went wrong. However, when the system is taken back to base to be examined by engineers and troubleshooters, a transparent system (or one that is at least *more*

---

[36] US Department of Defense (2023), p. 6, emphasis added.

[37] See, e.g., Adadi and Berrada (2018); Arrieta et al. (2020); Cambria et al. (2023); Das and Rad (2020); Fiok et al. (2021); Langer et al. (2021); Miller (2019); Peters (2022); Speith (2022).

[38] See, e.g., Ross (2024); Rudin (2019); Wood (2024a).

transparent) allows for rapid assessment and addressing of underlying issues. The more a system can provide clear explanations of its reasoning or the more its reasoning can be laid bare to engineers, the better they can modify and improve systems to meet needs or evolving challenges on the battlefield. This is not to say that traceability is a necessity, but it is certainly a virtue we ignore to our detriment.

In closing, it is also worth highlighting that the need for traceability may be context- or mission-dependent. The level of transparency and traceability needed for effective and responsible use within some deployment context will be contingent on the realities of that deployment context. Concretely, AWS or AI-enabled systems deployed to less morally or legally complex environments may not need to be as traceable, given that there will be fewer risks overall and combatants, therefore, need not necessarily have such an intimate or deep understanding of a system's underlying processes. For example, in aerial or naval environments, it may often be the case that one can know with relative certainty that only military assets will be in the area—ruling out mistakes of targeting—and the autonomous and AI-enabled systems will almost certainly be targeting assets or materiel rather than persons—largely ruling out mistakes relating to soldiers who are surrendering or *hors de combat*. In such environments, the core concern will be whether the system generally functions as required and can be trusted *in that specific environment*, but there is unlikely to be a strong need for traceability. That being said, traceability may still hold value, insofar as one may wish for systems that underperform or fail to achieve the mission to be examined to see whether their programming could be improved. Thus, it is possible that the value of effectiveness may still indicate in favor of at least some level of traceability. The overall lesson, though, is that traceability, like every design desideratum, must be pursued not just with our deeper moral, legal, social, and martial values in mind but also with an eye to the realities of warfare and what those realities imply for weapons design and use.

6.5 Technology Life-Cycles

All of the design choices and design desiderata examined in the previous section bear clear connection points. Designing for and improving predictability can lead to systems being more governable overall, making it easier to utilize systems and assign responsibility in case of mistakes. Likewise, improvements in governability may further predictability, or training regimens and military doctrine aimed at anchoring responsibility may make combatants more competent overall, leading to systems' outcomes being more predictable in real combat environments. Behind all of these, the traceability of systems is almost certain to feed into all other desiderata as well. In short, nothing is really to be considered in isolation. Nor should values at the design stage be viewed in isolation from deeper norms and principles, as selected values lead to design desiderata, desiderata lead to discrete design choices focusing on how systems will be used in real environments, and those design choices often further multiple values in synergistic and interconnected ways. Beyond what we may think of as the synchronic connections between design choices and values at a given time, there are also diachronic connections between technologies, values, and design choices *through* time.

Technologies can persist for long periods, and they bear our values long after we are done with them. The technologies we create and how exactly we design them can also shape our future values, closing off particular avenues of exploration and social organization, and sometimes they will do this in ways society would not want now and, in fact, might not want in the future and would prevent if we could go back and do things differently. These facts must be borne in mind here and now as we decide how best to design systems, ensuring that they represent and further our values here and now and that they will do so *throughout their life-cycle*.[39] When we consider the whole life cycle of autonomous and AI-enabled systems in the military, three main concerns arise.

First and most obviously, we must ask ourselves how, after we are done with AWS and military AI, we can secure these systems to prevent them from falling into the wrong hands. Across the literature, this is regularly considered a grave concern, with the proliferation of AWS to rogue states and non-state actors a recurring topic.[40] Unfortunately, it is a problem that likely does not have a solution. The most concerning forms of AWS, and those we would most wish to prevent rogue states and non-state actors from possessing, are lethal anti-personnel systems. Yet anyone with access to simple over-the-counter drones, small arms, and rudimentary computer programs can create these independently. Moreover, the less scrupulous an actor is—and hence the more we would want to prevent proliferation to them—the more easily they can cobble together simple (and likely indiscriminate) AWS. Most autonomous and AI-enabled systems being developed by advanced military powers, on the other hand, are high-end systems most often designed for defensive or anti-materiel purposes. Restrictions on these or efforts to prevent or limit proliferation will thus not significantly reduce the most egregious forms of AWS that might be out there.[41] Moreover, there is less risk associated with proliferation of, say, anti-tank AWS as opposed to anti-personnel AWS, but it is the former that we can more easily constrain. That being said, there are ways we can potentially limit the proliferation of the systems we are creating, even if that will not impact bad actors' ability to create more rudimentary systems on their own. Institutionally, arms controls and reductions would help to lower risks of proliferation. From the design perspective, we may demand that AWS and military AI be designed with hard-coded inbuilt deactivation protocols, such that systems will become unusable after some set

[39] These points are richly explored in the "Value Sensitive Design" framework. See, e.g., de Reuver et al. (2020); Iversen et al. (2020); Lavi and Reich (2024); Umbrello (2021b); Umbrello and van de Poel (2021).

[40] E.g., Altmann and Sauer (2017); Human Rights Watch (2016b, 2018); Kwik (2023); Sauer (2020); Zając (2022b).

[41] See Kwik (2024) for similar arguments.

period post-deployment.[42] While this expiration feature would not eliminate proliferation risks, it would mitigate them to some extent. And when this design decision is coupled with institutional and other approaches, it could address this concern.

A second worry is that the deployment and normalization of AWS and military AI may alter warfare in unwanted ways, changing not just how wars are fought but also the norms of war and what we expect of upright nations and combatants. To this, it is important to stress the potential magnitude of this concern and highlight that it is one we can likely never fully address, regardless of the domain in question. The simple fact is that technologies change; changing technologies can change our values, and it is often difficult to see with any confidence how our values ultimately will change. Even so, if we are designing autonomous and AI-enabled systems *with* and *for* our current martial values, values which have been a part of war for millennia and are unlikely to change rapidly, then we place ourselves in a strong position to be aware of this risk and be vigilant in meeting it as it may arise. Moreover, the more AWS and military AI are designed to entrench and promote human judgment, the more we can mitigate concerns surrounding dehumanization in war[43] or worries that AWS will make our combatants and political leaders less respecting of life in general.[44]

Finally, we have been focusing on autonomous and AI-enabled weapons used in war. However, if we are to take seriously the entire life-cycle of these technologies, we must also consider how purely military developments may be kept from infiltrating the civil sphere through policing, public surveillance, or intelligence operations. The simple fact is that we may design systems to give our states an advantage over foreign adversaries, but that advantage can also be turned on populations at home. This is especially relevant given how much AI has been harnessed in the military purely for intelligence gathering and processing purposes. Collecting and evaluating massive troves of data has clear implications for policing, distribution of public services, and many other elements of domestic governance. This cautionary tale is not to say that "weapons of war" will necessarily find their way onto the streets at home, but the underlying technologies for autonomous and AI-enabled weapons will clearly have relevance for everyday governmental operations. What exactly that means for the design of AWS and military AI is uncertain to me at this point, but it should be clear that responsible and value-conscious designers ought to bear this concern in mind when developing these technologies, and militaries ought to consider how the procurement of a system may have long-term trickle-down effects on the societies they are sworn to protect.

[42] This could follow the design decisions and reasoning of the United States military with regards to mine design, where mines have both an automatic deactivation timer (usually after only a few hours, or at most days) as well as a battery powering the trigger which has only a short life (usually only a couple weeks).

[43] E.g., Asaro (2012); Human Rights Watch (2012, 2018); Renic and Schwarz (2023); Stop Killer Robots (2022); Wagner (2014).

[44] E.g., Altman and Wellman (2008); Gubrud (2014); Masood and Baig (2023); Rickli (2015); Sharkey (2012). Cf. Wood (2022); Zając (2022a), Ch. 9.1 for responses to many of these arguments.

References

Adadi, A., & Berrada, M. (2018). Peeking inside the black-box: A survey on explainable artificial intelligence (XAI). *IEEE Access, 6*, 52138–52160.

Altman, A., & Wellman, C. H. (2008). From humanitarian intervention to assassination: Human rights and political violence. *Ethics, 118*(2), 228–257.

Altmann, J., & Sauer, F. (2017). Autonomous weapon systems and strategic stability. *Survival, 59*(5), 117–142.

Arendt, H. (1970). *On violence*. Harvest/HBJ.

Arrieta, A. B., Díaz-Rodríguez, N., Del Ser, J., Bennetot, A., Tabik, S., Barbado, A., Garcia, S., Gil-Lopez, S., Molina, D., Benjamins, R., Chatila, R., & Herrera, F. (2020). Explainable artificial intelligence (XAI): Concepts, taxonomies, opportunities and challenges toward responsible AI. *Information Fusion, 58*, 82–115.

Asaro, P. (2012). On banning autonomous weapon systems: Human rights, automation, and the dehumanization of lethal decision-making. *International Review of the Red Cross, 94*(886), 687–709.

Asaro, P. M. (2006). What should we want from a robot ethic? *International Review of Information Ethics, 6*, 9–16.

Bellamy, A. J. (2006). *Just Wars: From Cicero to Iraq*. Polity Press.

Blanchard, A., & Taddeo, M. (2022). Predictability, distinction & due care in the use of lethal autonomous weapon systems. *SSRN Electronic Journal*. https://doi.org/10.2139/ssrn.4099394

Bode, I., & Watts, T. F. (2021). *Meaning-less human control: Lessons from air defence systems on meaningful human control for the debate on AWS*. Technical report, Center for War Studies.

Boothby, W. H. (2016). *Weapons and the law of armed conflict* (2nd ed.). Oxford University Press.

Boshuijzen-van Burken, C. (2023). Value sensitive design for autonomous weapon systems—A primer. *Ethics and Information Technology, 25*(1).

Boulanin, V., Davison, N., Goussac, N., & Carlsson, M. P. (2020). *Limits on autonomy in weapon systems: Identifying practical elements of human control*. Technical report, International Committee of the Red Cross and Stockholm International Peace Research Institute.

Brunstetter, D. R., & O'Driscoll, C. (Eds.). (2018). *Just War thinkers: From Cicero to the 21st century*. Routledge.

Cambria, E., Malandri, L., Mercorio, F., Mezzanzanica, M., & Nobani, N. (2023). A survey on XAI and natural language explanations. *Information Processing & Management, 60*(1), 1–16.

Cappuccio, M. L., Galliott, J. C., & Sandoval, E. B. (2021). Saving private robot: Risks and advantages of anthropomorphism in agent-soldier teams. *International Journal of Social Robotics, 14*(10), 2135–2148.

Caron, J.-F. (2020). Defining semi-autonomous, automated and autonomous weapon systems in order to understand their ethical challenges. *Digital War, 1*(1–3), 173–177.

Chengeta, T. (2016). Accountability gap: Autonomous weapon systems and modes of responsibility in international law. *Denver Journal of International Law & Policy, 45*, 1–50.

Cox, R. (2018). The ethics of war up to Thomas Aquinas. In S. Lazar & H. Frowe (Eds.), *The Oxford handbook of ethics of war* (pp. 99–121). Oxford University Press.

Crootof, R. (2016). War torts: Accountability for autonomous weapons. *University of Pennsylvania Law Review, 164*, 1347–1402.

Crootof, R., Kaminski, M. E., & Price, W. N., II. (2023). Humans in the loop. *Vanderbilt Law Review, 76*, 429–510.

Daoust, I. (2001). ICRC expert meeting on legal reviews of weapons and the SIrUS project. *International Review of the Red Cross, 83*(842), 539–542.

Das, A., & Rad, P. (2020). Opportunities and challenges in explainable artificial intelligence (XAI): A survey. *arXiv*. https://doi.org/10.48550/arXiv.2006.11371

de Reuver, M., van Wynsberghe, A., Janssen, M., & van de Poel, I. (2020). Digital platforms and responsible innovation: Expanding value sensitive design to overcome ontological uncertainty. *Ethics and Information Technology, 22*(3), 257–267.

Dexe, J., Franke, U., Nöu, A. A., & Rad, A. (2020). Towards increased transparency with value sensitive design. In *International conference on human-computer interaction* (pp. 3–15). Springer.
Dige, M. (2017). Explaining the principle of mala in se. In D. Brown (Ed.), *The Just War tradition: Applying old ethics to new problems* (pp. 141–155). Routledge.
Dill, J., & Shue, H. (2012). Limiting the killing in war: Military necessity and the St. Petersburg assumption. *Ethics & International Affairs, 26*(3), 311–333.
Ekelhof, M. (2019). Moving beyond semantics on autonomous weapons: Meaningful human control in operation. *Global Policy, 10*(3), 343–348.
Fiok, K., Farahani, F. V., Karwowski, W., & Ahram, T. (2021). Explainable artificial intelligence for education and training. *The Journal of Defense Modeling and Simulation: Applications, Methodology, Technology, 19*(2), 133–144.
Friedman, B. (1996). Value-sensitive design. *Interactions, 3*(6), 16–23.
Friedman, B., & Hendry, D. G. (2019). *Value sensitive design: Shaping technology with moral imagination*. MIT Press.
Friedman, B., Hendry, D. G., & Borning, A. (2017). A survey of value sensitive design methods. *Foundations and Trends in Human-Computer Interaction, 11*(2), 63–125.
Friedman, B., Kahn, P., & Borning, A. (2002). Value sensitive design: Theory and methods. *University of Washington Technical Report, 2*(8), 1–8.
Friedman, B., Kahn, P. H., Borning, A., & Huldtgren, A. (2013). Value sensitive design and information systems. In N. Doorn, D. Schuurbiers, I. van de Poel, & M. E. Gorman (Eds.), *Early engagement and new technologies: Opening up the laboratory* (pp. 55–95). Springer.
Gardam, J. (2004). *Necessity, proportionality and the use of force by states*. Cambridge University Press.
Goodman, R. (2013). The power to kill or capture enemy combatants. *European Journal of International Law, 24*(3), 819–853.
Griffin-Heslin, V. L. (2005). An analysis of the concept dignity. *Accident and Emergency Nursing, 13*(4), 251–257.
Gubrud, M. (2014). Stopping killer robots. *Bulletin of the Atomic Scientists, 70*(1), 32–42.
Haines, S. (2014). The developing law of weapons: Humanity, distinction and precautions in attack. In A. Clapham & P. Gaeta (Eds.), *The Oxford handbook of international law in armed conflict* (pp. 273–295). Oxford University Press.
Haque, A. A. (2018). Necessity and proportionality in international law. In L. May (Ed.), *The Cambridge handbook of the Just War* (pp. 255–272). Cambridge University Press.
Heins, J. C. (2018). *Letting go of the loop: Coming to grips with autonomous decision-making in military operations*. Technical report.
Heller, K. J. (2023). The concept of "the human" in the critique of autonomous weapons. *Harvard National Security Journal, 15*(1), 1–76.
Hindriks, F., & Veluwenkamp, H. (2023). The risks of autonomous machines: From responsibility gaps to control gaps. *Synthese, 201*(1), 1–17.
Human Rights Watch. (2012). *Losing humanity: The case against killer robots*. Technical report, Human Rights Watch.
Human Rights Watch. (2015). *Mind the gap: The lack of accountability for killer robots*. Technical report, Human Rights Watch.
Human Rights Watch. (2016a). *Killer robots and the concept of meaningful human control*. Technical report, Human Rights Watch.
Human Rights Watch. (2016b). *Making the case: The dangers of killer robots and the need for a preemptive ban*. Technical report, Human Rights Watch.
Human Rights Watch. (2018). *Heed the call: A moral and legal imperative to ban killer robots*. Technical report, Human Rights Watch.
Hurka, T. (2005). Proportionality in the morality of war. *Philosophy & Public Affairs, 33*(1), 34–66.
Hurka, T. (2008). Proportionality and necessity. In L. May (Ed.), *War: Essays in political philosophy* (pp. 127–144). Cambridge University Press.

International Committee of the Red Cross. (1997). *The SIrUS project*. Technical report, International Committee of the Red Cross.

International Committee of the Red Cross. (2014). *Autonomous weapons systems: Technical, military, legal and humanitarian aspects*. Technical report, International Committee of the Red Cross.

International Committee of the Red Cross. (2021a). *ICRC position and background paper on autonomous weapons systems*. Technical report, International Committee of the Red Cross.

International Committee of the Red Cross. (2021b). *ICRC position on autonomous weapons systems*. Technical report, International Committee of the Red Cross.

Ivanov, S. H. (2023). Automated decision-making. *Foresight, 25*(1), 4–19.

Iversen, N., Birkved, M., & Cawthorne, D. (2020). Value sensitive design and environmental impact potential assessment for enhanced sustainability in unmanned aerial systems. In *2020 IEEE international symposium on technology and society (ISTAS)* (pp. 192–200). IEEE.

Johnson, J. (2023a). *The AI commander: Centaur teaming, command, and ethical dilemmas*. Oxford University Press.

Johnson, J. (2023b). Automating the OODA loop in the age of intelligent machines: Reaffirming the role of humans in command-and-control decision-making in the digital age. *Defence Studies, 23*(1), 43–67.

Johnson, J. (2024). Finding AI faces in the moon and armies in the clouds: Anthropomorphising artificial intelligence in military human-machine interactions. *Global Society, 38*(1), 67–82.

Kalmanovitz, P. (2018). Early modern sources of the regular war tradition. In S. Lazar & H. Frowe (Eds.), *The Oxford handbook of ethics of war* (pp. 145–166). Oxford University Press.

Kelsay, J. (2007). *Arguing the Just War in Islam*. Harvard University Press.

Killmister, S. (2010). Dignity: Not such a useless concept. *Journal of Medical Ethics, 36*(3), 160–164.

Königs, P. (2022). Artificial intelligence and responsibility gaps: What is the problem? *Ethics and Information Technology, 24*(3).

Kwik, J. (2023). Mitigating the risk of autonomous weapon misuse by insurgent groups. *Laws, 12*(1), 1–25.

Kwik, J. (2024). Desirability of using AWS. In *Lawfully using autonomous weapon technologies* (pp. 275–296). Springer.

Langer, M., Oster, D., Speith, T., Hermanns, H., Kästner, L., Schmidt, E., Sesing, A., & Baum, K. (2021). What do we want from explainable artificial intelligence (XAI)?—A stakeholder perspective on XAI and a conceptual model guiding interdisciplinary XAI research. *Artificial Intelligence, 296*, 1–24.

Lavi, E., & Reich, Y. (2024). Cross-disciplinary system value overview towards value-oriented design. *Research in Engineering Design, 35*(1), 1–20.

Lazar, S. (2012). Necessity in self-defense and war. *Philosophy & Public Affairs, 40*(1), 3–44.

Lazar, S. (2013). Necessity and non-combatant immunity. *Review of International Studies*, 1–24.

Macak, C. A. (2005). *Centaurs for maneuver warfare: Human-machine collaboration and manned-unmanned teaming for the fifth-generation ground combat element*. Technical report, United States Marine Corps Command and Staff College.

Masood, M., & Baig, M. A. (2023). Potential impact of lethal autonomous weapon systems on strategic stability and nuclear deterrence in South Asia. *Margalla Papers, 27*(2), 27–43.

Matthias, A. (2004). The responsibility gap: Ascribing responsibility for the actions of learning automata. *Ethics and Information Technology, 6*(3), 175–183.

Mayer, C. (2013). Minimizing harm to combatants: Nonlethal weapons, combatants' rights, and state responsibility. In F. Allhoff, N. G. Evans, & A. Henschke (Eds.), *Routledge handbook of ethics and war: Just War theory in the 21st century* (pp. 301–311). Routledge.

McMahan, J. (2013). Proportionate defense. *Journal of Transnational Law & Policy, 23*, 1–36.

McMahan, J. (2015). Proportionality and time. *Ethics, 125*(3), 696–719.

McMahan, J. (2018). Proportionality and necessity in Jus in Bello. In S. Lazar & H. Frowe (Eds.), *The Oxford handbook of ethics of war* (pp. 418–439). Oxford University Press.

Michael, L. (2014). Defining dignity and its place in human rights. *The New Bioethics, 20*(1), 12–34.
Miller, T. (2019). Explanation in artificial intelligence: Insights from the social sciences. *Artificial Intelligence, 267*, 1–38.
Müller, V. C. (2016). Autonomous killer robots are probably good news. In E. Di Nucci & F. Santoni de Sio (Eds.), *Drones and responsibility: Legal, philosophical and socio-technical perspectives on the use of remotely controlled weapons* (pp. 67–81). Routledge.
Neads, A., Galbreath, D. J., & Farrell, T. (2021). *From tools to teammates: Human-machine teaming and the future of command and control in the Australian Army*. Technical report, Australian Army Research Centre.
Nordenfelt, L. (2009). The concept of dignity. In *Dignity in care for older people* (pp. 26–53). Wiley.
Ohlin, J. D. (2012). The duty to capture. *Minnesota Law Review, 97*, 1268–1342.
Ohlin, J. D., & May, L. (2016). *Necessity in International Law*. Oxford University Press.
Oimann, A.-K. (2023a). The responsibility gap and LAWS: A critical mapping of the debate. *Philosophy & Technology, 36*(3).
Oimann, A.-K. (2023b). Why command responsibility may (not) be a solution to address responsibility gaps in LAWS. *Criminal Law and Philosophy, 18*(3), 765–791.
Onishi, K. (2018). Rethinking the permissive function of military necessity in internal non-international armed conflict. *Israel Law Review, 51*(2), 235–259.
Pacholska, M. (2023). Military artificial intelligence and the principle of distinction: A state responsibility perspective. *Israel Law Review, 56*(1), 3–23.
Pacholska, M. (2024). Autonomous weapons. In B. Brożek, O. Kanevskaia, & P. Pałka (Eds.), *Research handbook on law and technology* (pp. 392–407). Edward Elgar.
Pagallo, U. (2011). Robots of just war: A legal perspective. *Philosophy & Technology, 24*(3), 307–323.
Peters, U. (2022). Explainable AI lacks regulative reasons: Why AI and human decision-making are not equally opaque. *AI and Ethics, 3*(3), 1–12.
Pettit, P. (1997). *Republicanism: A theory of freedom and government*. Oxford University Press.
Reichberg, G. M. (2018). Historiography of Just War theory. In S. Lazar & H. Frowe (Eds.), *The Oxford handbook of ethics of war* (pp. 59–79). Oxford University Press.
Reijers, W., & Gordijn, B. (2019). Moving from value sensitive design to virtuous practice design. *Journal of Information, Communication, and Ethics in Society, 17*(2), 196–209.
Renic, N., & Schwarz, E. (2023). Crimes of dispassion: Autonomous weapons and the moral challenge of systematic killing. *Ethics & International Affairs, 37*(3), 321–343.
Rickli, J.-M. (2015). *Some considerations on the impact of LAWS on international security: Strategic stability, non-state actors and future prospects*. Technical report, Meeting of Experts on Lethal Autonomous Weapons Systems: Convention on Certain Conventional Weapons (CCW).
Roff, H. (2024). Magnifying human confusion: Meaningful human control and the ongoing debate on autonomous weapons. *The Rule of Law Post*. https://www.penncerl.org/the-rule-of-law-post/magnifying-humanconfusion-meaningful-human-control-and-the-ongoing-debate-onautonomous-weapons/
Roff, H. M., & Danks, D. (2018). "Trust but verify": The difficulty of trusting autonomous weapons systems. *Journal of Military Ethics, 17*(1), 2–20.
Rosen, M. (2012). *Dignity: Its history and meaning*. Harvard University Press.
Rosert, E., & Sauer, F. (2019). Prohibiting autonomous weapons: Put human dignity first. *Global Policy, 10*(3), 370–375.
Ross, A. (2024). AI and the expert; a blueprint for the ethical use of opaque AI. *AI & Society, 39*(3), 925–936.
Rudin, C. (2019). Stop explaining black box machine learning models for high stakes decisions and use interpretable models instead. *Nature Machine Intelligence, 1*(5), 206–215.
Saket, M. (2021). Putting values in context: An augmentation of value sensitive design (VSD). *Journal of Ethics and Emerging Technologies, 31*(2), 1–9.
Sancar, I. V. (2024). How can we design autonomous weapon systems? *AI and Ethics*, 1–9.

Sandoz, Y., Swinarski, C., & Zimmermann, B. (Eds.). (1987). *Commentary on the additional protocols*. Martinus Nijhoff.
Santoni de Sio, F., & van den Hoven, J. (2018). Meaningful human control over autonomous systems: A philosophical account. *Frontiers in Robotics and AI, 5*, 1–14.
Sauer, F. (2020). Stepping back from the brink: Why multilateral regulation of autonomy in weapons systems is difficult, yet imperative and feasible. *International Review of the Red Cross, 102*(913), 235–259.
Scharre, P. (2016). Centaur warfighting: The false choice of humans vs. automation. *Temple International and Comparative Law Journal, 30*, 151–165.
Schwartz, D. (2018). Late scholastic just war theory. In S. Lazar & H. Frowe (Eds.), *The Oxford handbook of ethics of war* (pp. 122–144). Oxford University Press.
Sharkey, N. (2007). Automated killers and the computing profession. *Computer, 40*(11), 124–123.
Sharkey, N. (2012). Killing made easy: From joysticks to politics. In P. Lin, K. Abney, & G. A. Bekey (Eds.), *Robot ethics: The ethics and social implications of robotics* (pp. 111–128). MIT Press.
Skinner, Q. (2002). A third concept of liberty. *Proceedings of the British Academy, 117*, 237–268.
Skinner, Q. (2008). Freedom as the absence of arbitrary power. In C. Laborde & J. Maynor (Eds.), *Republicanism and political theory* (pp. 83–101). Blackwell.
Solis, G. D. (2016). *The law of armed conflict: International humanitarian law in war* (2nd ed.). Cambridge University Press.
Sorabji, R., & Rodin, D. (Eds.). (2006). *The ethics of war: Shared problems in different traditions*. Ashgate.
Sparrow, R. (2007). Killer robots. *Journal of Applied Philosophy, 24*(1), 62–77.
Speith, T. (2022). A review of taxonomies of explainable artificial intelligence (XAI) methods. In *Proceedings of the 2022 ACM conference on fairness, accountability, and transparency* (pp. 2239–2250).
Statman, D. (2011). Can wars be fought justly? the necessity condition put to the test. *Journal of Moral Philosophy, 8*(3), 435–451.
Steinhoff, U. (2017). Proportionality in self-defense. *The Journal of Ethics, 21*(3), 263–289.
Stop Killer Robots. (2018). Facing fully autonomous weapons. Retrieved October 25, 2023, from https://www.youtube.com/watch?v=oTtoPCjBgxQ
Stop Killer Robots. (2022). *Negotiating a treaty on autonomous weapons systems: The way forward*. Technical report, Stop Killer Robots.
Taddeo, M., & Blanchard, A. (2022). A comparative analysis of the definitions of autonomous weapons systems. *Science and Engineering Ethics, 28*(5), 1–22.
Tomlin, P. (2019). Subjective proportionality. *Ethics, 129*(2), 254–283.
Tomlin, P. (2020). Proportionality in war: Revising revisionism. *Ethics, 131*(1), 34–61.
Umbrello, S. (2021a). Coupling levels of abstraction in understanding meaningful human control of autonomous weapons: A two-tiered approach. *Ethics and Information Technology*, 1–10.
Umbrello, S. (2021b). *Towards a value sensitive design framework for attaining meaningful human control over autonomous weapons systems*. PhD thesis, Consortium Fino.
Umbrello, S. (2022). *Designed for death: Controlling killer robots, Volume 1 of Ethics and robotics*. Trivent.
Umbrello, S., & van de Poel, I. (2021). Mapping value sensitive design onto AI for social good principles. *AI and Ethics, 1*(3), 283–296.
Uniacke, S. (2011). Proportionality and self-defense. *Law and Philosophy, 30*(3), 253–272.
US Army. (1956). *FM 27-10: The law of land warfare*. Technical report, United States Army.
US Army Judge Advocate General. (2015a). *Law of armed conflict deskbook*. Technical report, International and Operational Law Department, United States Army.
US Army Judge Advocate General. (2015b). *Operational law handbook*. Technical report, International and Operational Law Department, United States Army.
US Department of Defense. (2023). *DoD Directive 3000.09*. Technical report, United States Department of Defense.

van den Hoven, J. (2013). Value sensitive design and responsible innovation. In R. Owen, J. Bessant, & M. Heintz (Eds.), *Responsible Innovation: Managing the responsible emergence of science and innovation in society* (pp. 75–83). Wiley.

van den Hoven, J., & Manders-Huits, N. (2020). Value-sensitive design. In K. W. Miller & M. Taddeo (Eds.), *The ethics of information technologies* (pp. 329–332). Routledge.

Verdiesen, I., & Dignum, V. (2023). Value elicitation on a scenario of autonomous weapon system deployment: a qualitative study based on the value deliberation process. *AI and Ethics, 3*, 887–900.

Wagner, M. (2014). The dehumanization of international humanitarian law: Legal, ethical, and political implications of autonomous weapon systems. *Vanderbilt Journal of Transnational Law, 47*, 1371.

Williams, A. P. (2015). Defining autonomy in systems: Challenges and solutions. In A. P. Williams & P. D. Scharre (Eds.), *Autonomous systems: Issues for defense policymakers* (pp. 27–62). NATO Communications and Information Agency.

Winner, L. (1978). *Autonomous technology: Technics-out-of-control as a theme in political thought*. MIT Press.

Winner, L. (1980). Do artifacts have politics? *Daedalus, 109*(1), 121–136.

Winner, L. (1986). *The whale and the reactor: A search for limits in an age of high technology* (2nd ed.). University of Chicago Press.

Wood, N. G. (2022). Autonomous weapons systems and force short of war. *Journal of Ethics and Emerging Technology, 32*(2), 1–16.

Wood, N. G. (2023a). Autonomous weapon systems: A clarification. *Journal of Military Ethics, 22*, 1–15.

Wood, N. G. (2023b). Autonomous weapon systems and responsibility gaps: A taxonomy. *Ethics and Information Technology, 25*(1), 1–14.

Wood, N. G. (2024a). Explainable AI in the military domain. *Ethics and Information Technology, 26*(2), 1–13.

Wood, N. G. (2024b). Reliability standards for (autonomous) weapons. *Ethics and Armed Forces, 24*(1).

Wood, N. G. (forthcoming). AI, trust, and teaming: The humans-as-handlers approach for autonomous and opaque AI systems. *Minds and Machines*.

Wood, N. G. (2025a). Autonomous and AI-enabled systems: Extensions or replacements of human will and control? *Ethics and Information Technology*. https://doi.org/10.1007/s10676-025-09876-9

Wood, N. G. (2025b). Bombs, bots, and the principle of distinction: The law of armed conflict and contemporary warfare. *Texas National Security Review, 9*(1).

Wyatt, A., & Galliott, J. (2021). An empirical examination of the impact of cross-cultural perspectives on value sensitive design for autonomous systems. *Information, 12*(12), 527.

Zając, M. (2022a). *Autonomous weapon systems from a Just War theory perspective*. PhD thesis, University of Warsaw.

Zając, M. (2022b). Beyond deadlock; low hanging fruit and strict yet achievable options in AWS regulations. *Journal of Ethics and Emerging Technologies, 32*(2), 1–14.

Chapter 7
Technical Alienation, AI, and Autonomous Technology

Sergio Amadeu da Silveira

Abstract The purpose of this chapter is to present an interpretation of Langdon Winner's perspective on the problem of autonomous technology, which relates to the debate on the notion of alienation. I also intend to show here that by decoding the phenomenon of autonomous and alienated technology, we can more clearly observe the current processes of knowledge and power centralization that actually existing Artificial Intelligence is promoting. The theoretical tools brought by Winner in his two books, *Autonomous Technology* (1978) and *The Whale and the Reactor* (1986), allow us to observe the automated systems and their current technopolitical implications in depth. Despite having a different perspective, the application of the concept of alienation that Winner makes can align with Gilbert Simondon's critique of the separation between culture and technology. For Simondon, the distance between humans and technical objects leads to a lack of understanding of technicality, a phenomenon that, through the prism of AI, is exacerbated by corporate structures that hold a monopoly over technological development. This distance prevents individuals and society from understanding the cultural, social, and political implications of the technologies that shape their lives. When analyzing this issue, it becomes clear that actually existing AI not only reproduces existing inequalities but also reconfigures society to maintain and expand these power structures.

Keywords Technopolitics · Autonomous technology · Technological alienation

S. A. da Silveira (✉)
CECS, Universidade Federal do ABC, São Bernardo do Campo, SP, Brazil
e-mail: sergio.amadeu@ufabc.edu.br

© The Author(s), under exclusive license to Springer Nature Switzerland AG 2026

A. Fox, D. Cressman (eds.), *Arrangements of Power: Tracing Langdon Winner's Legacy Within and Beyond the Philosophy of Technology*, Philosophy of Engineering and Technology 50, https://doi.org/10.1007/978-3-032-17371-3_7

7.1 Introduction

The contributions of Langdon Winner on the technological systems that drive and govern contemporary life are essential for a deep understanding of the social, economic, and political implications of "actually-existing AI", a term coined by Dyer--Witheford et al. (2019) to demystify automated systems. In his book *Autonomous Technology: Technics-out-of-Control as a Theme in Political Thought*, published in 1978, Winner suggests that certain technologies often seem to operate outside human control, surpassing and altering the purposes and modes of use for which they were created.

The complexity of technologies, the increasing specialization, and specific technical imperatives are pointed out by Langdon Winner to demonstrate the tendency of complex technical systems to follow paths independently of human will: once created and set in motion, these technical systems seem to acquire their own dynamics. As technical systems become increasingly sophisticated, they make it impossible for a single individual or group to understand and control all of their components. Specialization fragments knowledge, leading to its concentration among specialists and corporations.

In the book *The Whale and the Reactor*, Winner (1986) deepens this perception by bringing the perspective of knowledge concentration along with that of alienation, generating a *technological somnambulism*:

> Artifacts in common use presuppose relatively few centers of production and distribution. In addition, many of our daily activities rely upon systems that we do not make, control, or know how to repair when they break down. It is possible to imagine alternatives to this state of affairs, different ways of structuring the physical and social dimensions of modern material culture. But given the direction that technological "progress" has followed, people find themselves dependent upon a great many large, complex systems whose centers are, for all practical purposes, beyond their power to influence. (1986, p. 5)

The purpose of this chapter is to present an interpretation of Langdon Winner's perspective on the problem of autonomous technology, which relates to the debate on the notion of alienation. I also intend to show here that by decoding the phenomenon of autonomous and alienated technology, we can more clearly observe the current processes of centralization and concentration of knowledge and power that actually existing Artificial Intelligence is promoting. The theoretical tools brought by Winner in his two books, *Autonomous Technology* (1978) and *The Whale and the Reactor* (1986), allow us to observe the automated systems and their current technopolitical implications in depth.

7.2 Alienation in Marx, Simondon, and Winner

We can distinguish Langdon Winner's approach to alienation from those made by Karl Marx and Gilbert Simondon, two thinkers who employed it in significant ways. Winner did not provide his own definition of "alienation," but this concept is implicit in his reflections on technology and power. Alienation is a social process and has a direct relationship with individual and collective consciousness. It manifests in common sense and doctrines such as positivism that preach the idea of socially neutral technology. But not only that. Technical alienation reinforces the prevailing power of large technology companies, commonly referred to as Big Techs (Morozov, 2018) and consolidates the processes of concentrating the essential inputs of actually existing AI into a few infrastructures, deepening asymmetries and the concentration of power in the current international division of labor and knowledge.

The Marxist approach to alienation is centered on labor. For Marx (1971), the worker is alienated from the product of his work, from the production process, and thus from his own human essence. Alienation is a direct consequence of capital's exploitation and control over labor. It is structured by the private ownership of the means of production and the division of the production process. In capitalism, the worker becomes a cog in the machine of mass production, with no control over the process or the final product.

Marx stated that the development of productive forces by introducing machines and automated systems turned the worker into a mere extension of the machine. His main function is no longer the exercise of his creative and productive abilities but the operation of machines that dictate the rhythm, functions, and content of his activity. In this context, the worker does not control the production process but is controlled by it. The machine determines the pace of work, limiting human labor to overseeing or adjusting a technical process. Thus, the worker is alienated and unable to recognize himself in the product of his own labor.

Gilbert Simondon's (1980) perspective on alienation is quite different. While Marx considered alienation embedded and originating in economic and social relations, especially regarding labor, the production of commodities, and class structure, Simondon placed it in the relationship between technology and the human being, emphasizing the importance of overcoming alienation through the understanding of technical objects and the reintegration of technology into culture. Thus, Simondon defines alienation in the context of the separation between humans and technical objects, between culture and technology. For him, this separation occurs because society does not understand the technicality and potential of technical objects. Alienation arises from the lack of integration between humans and technology.

Simondon criticized Marx for not taking into account the 'psychophysiological' alienation of workers during the machine age (Barthélémy, 2012). For Simondon, modern technical development, by emphasizing technical rationality at the expense of bodily and emotional needs, has led to a form of psychophysiological alienation. This also connects with his broader concept of technical alienation, where the separation between humans and technical objects results in a distancing from both technical understanding and the social and personal implications of these technologies. By isolating the mind from the body, psychophysiological alienation exacerbates the feeling of lack of control or understanding over the technical world and its consequences on human life.

> The fundamental stake of MEOT [*On the Mode of Existence of Technical Objects*] is to reconcile culture with technics by supporting the introduction of a 'technical culture,' which is necessary today for the very equilibrium of culture: 'Culture must become general again, whereas now it is specialized and impoverished. Such an extension of culture, which would suppress one of the principle sources of alienation and would reestablish regulative information, has political and social value' (MEOT 14). Culture is defined as: 'that by which the human being regulates its relation to the world and its relation to itself' (MEOT 227). In order fundamentally to reconcile culture with technics, Simondon will embark in MEOT on a complex operation that consists in reconciling nature simultaneously and to an equal extent with both culture and technics. (Barthélémy, 2012, p. 209)

For Simondon, technical alienation arises when humans fail to recognize the process of technical individuation, that is, the way technical objects evolve, adapt, and interact with their environment. Simondon argues that technology is relational by nature, and technical objects have a complex genesis consisting of interactions between matter, energy, and information. By fragmenting technology from its genesis and treating it merely as a final product, modern society obscures it as inseparable from humanity and its culture. Technology is one of the greatest cultural expressions of societies and their creative drive. The absence of this perception is one of the most elementary characteristics of technical alienation.

In Marx's view, the revolutionary overthrow of capitalism could begin the journey toward overcoming labor alienation and freeing the worker from subordination to machines. For Simondon, overcoming labor alienation would not eliminate technical alienation. This alienation can only be overcome through understanding the mode of existence of technical objects and their process of concretization, which is the evolutionary development of technologies, where their functions increasingly integrate into our lives and culture. Therefore, Simondon argues that humanity must understand how machines work and how they relate to the social and cultural environment as an extension of their own inventive capacity and not as something foreign, alien, and alienating. However, in Simondon's perspective, the relationship between technical alienation and power (and the appropriation by social groups that structure the subordination of other human groups through technologies) is secondary. This greatly distances Winner's perspective from Simondon's proposal.

In *Autonomous Technology*, Winner (1978) discusses alienation concerning the issue of autonomous technology. Winner's reading of Marx raises the question of whether the master, the elite, or the owners of machines and technologies could lose control over them, which is directly related to the question of power in society. To build his argument, Winner revisited the fictional writings of Edward Morgan Forster from the late 1920s to address the relationship of human-machine dependence. Instead of humanity being subordinated to mechanized and monotonous patterns, Forster's fiction was concerned with the rise of a pathological dependence that drains all activity and spirit from men. The life of the whole society is transferred into the mechanism.

Currently, this technological mediation is noticeable when we observe the dominance of various digital platforms and their interfaces accessed through mobile phones. These platforms have become indispensable to the daily existence of new generations, whether through TikTok, Instagram, Snapchat, WhatsApp, or tools like ChatGPT, Gemini, Claude etc.

"The price paid for this wonder is that human beings can no longer have any direct relationship with each other or with the world; everything must go through the technical intermediary" (Winner, 1978, p. 35). This is an aspect of machine autonomy in its relationship with human autonomy, exposing subjugations that form subjects, which brings us back to the debate about social relations as relations of power.

Marx had already pointed out that machines absorbed or subsumed part of extended human activity through labor as self-activity, which is a substantial part of humanity's relationship with nature. Winner warned us that the German thinker preferred to call the loss of autonomy "alienation," evoking images of technological animism. Marx, when describing the consolidation of nineteenth-century capitalism, noted that technology had assumed a certain independence, as opposed to the worker, as a colossal and even monstrous force, thus alien:

> For Marx, the alienation of labor and the appearance of a massive, life-draining industrial mechanism cannot be considered independently of the historical class struggle, surplus value in industrial economics, capitalist accumulation, and the social and political domination of the bourgeoisie. Marx introduces the specter of an alien technological force to illustrate how far historical alienation has gone and how little power the worker has over his own daily existence. But Marx makes it clear that mastery over technology has not been really lost. It has simply been removed to a small segment of the social order, the capitalist class. Directly following the animistic passages of Capital quoted above, he stresses that there is most definitely a "master" "in whose brain the machinery and his monopoly of it are inseparably united." The individual worker is indeed dwarfed and diminished by his relationship to the technical means with which he has contact. But behind the mechanism stands a human figure who takes up and manipulates for his own advantage the power that has been drained from the proletariat. (Winner, 1978, 39–40)

Winner emphasized that, in Marx's theory, machines became autonomous in relation to workers and not in relation to their owners, who installed them in industries. While workers are subordinated to machines, the machinery is subjected to its owners, the capitalists. The alien power in Marx would be the appearance of the dominant class's power that employs the machine in production, subjecting the worker to its pace. The machine appears as if it were creating and producing within the alienating fog of labor in capitalist societies.

In *The Whale and the Reactor*, Winner (1986) explored alienation as an effect of individuals' distancing from the technological processes that shape their lives. This occurs when technologies are not just neutral tools but structures that shape social, political, and economic organizations. These technologies centralize control and decision-making, resulting in a disconnection between individuals and the forces shaping their daily lives.

In this context, Winner views alienation through the lens of technological decisions made by political or economic elites without the active participation of those most affected by these decisions. Technologies create social worlds to which people are subjected, forging subjects through systems they neither control nor fully understand. This intensifies the feeling of complete detachment from the decisions governing their lives. Winner linked this aspect of alienation to the notion of "technological somnambulism," which he used to refer to the tendency of societies to passively accept technological innovations without questioning their broader social implications.

7.3 The Dynamics of Concentration

Even in the industrial world, Winner had observed the tendency for the consumption and use of technologies to be widely decentralized while production and control became increasingly centralized. This tendency seems to stem from the capitalist system, which promotes economic concentration through the dominance of innovative technologies. One of Winner's major contributions was uncovering the alienation that emerges from this centralization and distancing from the poles of invention. Thus, Winner directed our attention to power relations and the tendency of industrial capitalism to reduce the number of technological centers "in which action is initiated and control is exercised".

> The tendency to remove the control over production from centers of everyday life—the home, neighborhood, community, and workplace—has become firmly fixed in both technical hardware and organizational practices of modern society. The designs of all kinds of technologies—electrical power plants, water systems, highways, and vehicles that run on them, the machines and chemicals of agribusiness farming, supermarkets and the products they stock, television, radio, computers, and many others—embody centralization in material form. Artifacts in common use presuppose relatively few centers of production and distribution. In addition, many of our daily activities rely upon systems that we do not make, control, or know how to repair when they break down. It is possible to imagine alternatives to this state of affairs, different ways of structuring the physical and social dimensions of modern material culture. But given the direction that technological "progress" has followed, people find themselves dependent upon a great many large, complex systems whose centers are, for all practical purposes, beyond their power to influence. (Winner, 1986, p. 94)

This centralizing propensity has intensified in information capitalism based on technologies for storing, processing, and handling information. With billions of people enmeshed in online social networks, trapped by attention-capturing triggers, and drawn into discount promotions for the latest versions of smartphones, the emergence of digital technologies and services (including generative AI applications) consolidates the common view of technological neutrality. The exciting technological convenience brought by Big Techs' common slogan, "we just want to improve your experience," reinforces the alienating idea of technological neutrality and the notion that technology is always good, and only its use might be problematic.

Winner's perspective, emphasizing political, organizational, and economic choices embedded in technologies, allows us to see how information technologies, especially actually existing Artificial Intelligence, have brought a centralizing and concentrating drive perhaps greater than that observed in industrial technologies.

Let us consider the implications of actually existing AI. The hegemonic approaches to AI, which are called machine learning and its variations, including deep learning, large language models (LLMs), and even generative AI, are based on extracting patterns from large amounts of data using enormous computational power. Data are the essential input for actually existing AI. Big Techs control the data and the data storage providers with their hyperscale data centers, known as cloud providers, which are indispensable infrastructures for training AI models. The so-called training of AI algorithms to refine their pattern-capturing abilities from databases requires substantial computational processing. The concentration of the generation of significant AI models in a few gigantic companies is alarming.

Large technology corporations have taken control of AI development despite the apparent democratization of AI development worldwide. There is a website called "There's an AI for That" that aims to gather tools, applications, and AI resources. As of early October 2024, the site had registered 18,680 AI projects created. This astounding number might be misleading, indicating a trend toward democratizing AI development. However, the opposite process is occurring: Large companies like Microsoft, Alphabet, Meta, Amazon, and IBM, in partnership with newer companies like OpenAI and Anthropic are concentrating on the development and training of large language models and other generative AI models. Generally, the recent strategy of large corporations was inspired by open-source software communities. Big Techs release open AI frameworks online. However, to run them, it usually requires a set of computers with high processing capacity. Startups and the general public end up using the computational infrastructure of corporations that offer it. Thus, small and medium institutions provide more data to the infrastructure of Big Techs, in addition to using large ready-made models to train their smaller models. By doing so, they improve the models of the large corporations that control the infrastructures.

The training of cutting-edge machine learning models has prohibitive costs for an ordinary person or even for a university in poor or middle-income countries. Furthermore, a single hyperscale data center exerts significant environmental impacts, requires a tremendous amount of energy, and consumes millions of gallons of water to cool and maintain the appropriate temperature for its thousands of servers. OpenAI's GPT-4, for instance, cost approximately US $78 million in computing to train its model (Maslej et al., 2024, p. 30). This was the cost of using Microsoft Azure cloud servers to extract patterns and fine-tune the GPT-4 model's parameters. Google's Gemini Ultra model consumed US $191 million to be trained (Maslej et al., 2024, p. 30). The increasing computing costs for generating and fine-tuning an actually existing AI model led to worrisome consequences, distancing not only ordinary citizens from understanding the political and doctrinal decisions adopted in its development but also the universities themselves.

As an example of this trend toward concentration, in 2024, Big Techs developed 51 notable models, i.e., sophisticated and highly complex models, while only 15 notable models were developed by universities. This trend of technoscientific leadership migrating from Academia to Big Techs has been ongoing since 2014. Until that year, academia led the release of machine learning models (Maslej et al., 2024, p. 46). This privatization of scientific research in a transversal and strategic field has occurred due to the hegemony of AI approaches based on processing vast amounts of data, which requires colossal storage and computational infrastructures, alienating research centers without abundant financial resources.

At this point, it is important to bring back Winner's debate on democratic and authoritarian technologies based on Lewis Mumford's works, both in *Autonomous Technology* (1978) and *The Whale and the Reactor* (1986). Mumford critiques authoritarian megatechnology. Winner reports that large-scale systems embody the choices of what Mumford called authoritarian megatechnics, which is preferred over small-scale systems, crafts, and trades that he named democratic polytechnics (Winner, 1978, p. 109). It would be hard to ignore that extant AI, especially deep learning models, could be seen as a megatechnic obsession seeking to control everything, as Mumford indicated. While free software communities develop their projects in a distributed manner, following the polytechnic tradition, Big Techs are voracious data collectors and builders of mega storage and processing structures, such as hyperscale data centers.

The side effects of this data concentration, which are the essential inputs for actually existing AI, are numerous—political, cultural, economic, and others. However, the least expected effect by digitalization optimists at the start of the twenty-first century is the disastrous environmental impact caused by data concentration in cloud providers and the current AI model. In light of this extreme concentration, it is necessary to consider the less apparent, deeper consequences that extend well beyond the technological domain and involve significant environmental costs, as illustrated by the following examples. Only Google's data center in the city of The Dalles, on the banks of the Oregon River in the United States, consumed approximately 355.1 million gallons of water in 2021 to cool and dissipate heat from the thousands of servers that store and process data for the technology giant (Rogoway, 2023). According to Columbia University, data centers today are responsible for 2.5–3.7% of global greenhouse gas emissions, surpassing the aviation industry (Cho, 2023). According to the International Energy Agency (IEA), the electricity consumption of data centers in 2022 was about 460 terawatt-hours (TWh), equivalent to France's annual energy consumption, and could reach over 1000 TWh by 2026, which would be equivalent to Japan's electricity consumption (IEA, 2024).

The most startling revelation is discovering that a single ChatGPT query requires 2.9 watt-hours of electricity, compared to 0.3 watt-hours for a Google search, according to the International Energy Agency (Goldman Sachs, 2024). This energy use is equivalent to the energy consumption of a 10 W LED bulb for 17 min. The business model based on the constant and ever-growing creation and collection of data seems to be environmentally disastrous and unsustainable. Technical alienation, which generates estrangement and social distancing from technologies, has contributed to societies, for at least a decade, failing to notice that the excessive and permanent accumulation of data would require colossal physical structures and a correspondingly massive energy expenditure at a time when our planet shows unequivocal signs of environmental collapse.

What initially appeared to be a democratizing technology, AI has proven to be highly centralizing. The controllers of significant and notable AI development concentrate AI's fundamental inputs (data), the storage, processing, and training infrastructures (cloud provisioning), and the frantic patenting of algorithms, models, and automated systems. We are dealing with technologies that have become highly detrimental to democracy, as well as to income and power concentration—what we can call an authoritarian megatechnology.

7.4 Automated Systems and Machines Out of Control

On July 19, 2024, Microsoft conducted a security update on the Windows operating system using the CrowdStrike Falcon solution, created and maintained by a third-party company specializing in cybersecurity. This solution, developed and maintained by CrowdStrike, was designed to detect and protect Windows software from cyberattacks, using "AI and machine learning to protect client systems by identifying and remediating the latest advanced threats" (O'Flaherty, 2024). Curiously, the application that was supposed to protect the system was responsible for shutting down all the machines that used it. The security solution affected many of Microsoft's client companies, including hospitals, schools, and airlines.

Philosopher Yuk Hui (2020) named similar events "algorithmic catastrophes" and warned us that they are not merely material failures but failures of reason. Inspired by Paul Virilio, Yuk Hui sees contemporary technological systems as carriers of catastrophes and techniques to mitigate the very tragedies that their dynamics and purposes generate. Catastrophes are inevitable due to the very nature of automation and automatization technologies. Thus, the administration of human activities is increasingly moving toward using machine intelligence solutions based on statistics and probability, converted into algorithmic systems that operate with immense computational power, generating models that are used to automate activities and mitigate the risks brought by automation itself.

When analyzed in terms of potential errors, biases, definitions, and unforeseen actions, deep learning algorithmic systems bring back Winner's (1986) warning that technical apparatuses can follow unexpected and unforeseen paths under various conditions, ultimately spiraling out of control. Given the high opacity of digital systems and their distance even from specialists within the companies and institutions using them, the trend is toward increasing the use of sensors and error detection devices, which will add even more complexity and the possibility of further mishaps. This alienating process is reinforced by the neoliberal doctrine that imposes, for example, on the public sector, a private-sector dynamic in which widely used algorithmic systems are developed and controlled by private companies that do not disclose their codes or trade secrets. Thus, the growing use of AI by democratic states consolidates increasingly opaque operations for society and even for public managers.

Acknowledging that, given the complexity, growing opacity, and gigantism of current technological operations, technologies can get out of control from their operators and even creators does not imply claiming that machines can acquire autonomy by exercising a will or asserting consciousness. It is about machinic agency: a capacity to act without conscious objectives. Machines, especially with the emergence of machine learning algorithms, are not creating new forms of knowledge by themselves but are automating social and cultural processes on a massive scale. This leads to a false impression of "autonomy" when, in fact, what we observe is an amplification of social control processes and algorithmic governance. Automation, especially with neural networks and large learning models, imitates and organizes forms of social cooperation but under a control paradigm, not human emancipation.

Matteo Pasquinelli (2023), in *The Eye of the Master*, argues that the idea of machine autonomy is, in fact, an ideological construction that masks the processes of control and centralization of human labor and social cooperation. For Pasquinelli, machine autonomy means that human control over labor and social life has been largely automated and centralized through AI systems. Pasquinelli sees technical autonomy as an extension of social control, where machines replicate and amplify the division of labor and hierarchies rather than acting independently or creatively. Moreover, emphasizing the control of human groups over technologies will not prevent the errors, biases, and catastrophes that digital technologies, particularly AI, may cause.

Winner (1978) discussed machine autonomy in a more literal sense, where technology can get out of human control and follow an unexpected trajectory. This does not seem to negate Pasquinelli's assertion that machines, especially the logical machines of AI, are deeply rooted in social and economic dynamics that control, rather than liberate, humanity. Nor can we dismiss Winner's perspective that this autonomy, understood as the loss of human control over technology, has become an emerging threat that must be carefully managed. It is reasonable to see the "autonomy" of actually existing AI more as a phenomenon of deep automation that consolidates the division of labor and amplifies social inequalities rather than genuine independence or creativity on the part of machines.

The idea of autonomous technology reinforces reification, as it exempts human and social responsibility for technical apparatuses. This reification was one of the main arguments presented by Winner in *Autonomous Technology* (1978). Additionally, the perspective of autonomy, presented as a certain consciousness or will of the machine, contributes to removing technological decisions from the political sphere. This perspective gained significant traction with AI and reinforced the idea of human subordination to machines. Such discourses about the existence or development of a strong and fully autonomous AI support and integrate the proposition of technology being detached from economic interests and impossible to subject to democratic control. This detachment only strengthens the power of Big Techs, the controllers of these automated systems, further distancing technoscientific inventions from societal understanding.

In this sense, societies become more alienated and increasingly distanced from the possibilities of dictating the direction of technologies, which are fundamental components of lifestyles and the structuring of power. It is important to recall what Winner (1986) called the "technological imperative," i.e., specific technologies that require the continuous adaptation of societies. Such imperatives lead to constantly reorganizing norms and practices while reinforcing the interdependence between technological progress and forms of political and economic power:

> If the experience of modern society shows us anything, however, it is that technologies are not merely aids to human activity, but also powerful forces acting to reshape that activity and its meaning. The introduction of a robot to an industrial workplace not only increases productivity but often radically changes the process of production, redefining what 'work' means in that setting. When a sophisticated new technique or instrument is adopted in medical practice, it transforms not only what doctors do but also the ways people think about health, sickness, and medical care. Widespread alterations of this kind in techniques of communication, transportation, manufacturing, agriculture, and the like are largely what distinguishes our times from early periods of human history. The kinds of things we are apt to see as 'mere' technological entities become much more interesting and problematic if we begin to observe how broadly they are involved in conditions of social and moral life. (Winner, 1986, p. 6)

Humans create technologies within their historical and social conditions, but these inventions materialized in artifacts have the power to reconfigure societies. In *The Eye of the Master*, Pasquinelli (2023) introduces the expression "the eye of the master" to address a historical and conceptual framework in which the role of the "master" was to supervise, control, and optimize the division of labor, initially in industrial contexts and later through digital automation. The metaphor of the "eye of the master" involves practices of surveillance and discipline imposed on workers, first through manual factory supervision and later through AI and algorithms in modern digital work. Thus, the "eye" of the master not only supervised but also disciplined workers by organizing and optimizing their movements, often stripping them of autonomy. As Winner indicated, technologies reorganize environments. In the current context, this idea extends to digital platforms and AI systems, where control and the division of labor are increasingly automated, centralizing power and control in the hands of a few. This metaphor connects to broader themes of surveillance, labor control, and the use of technology to perpetuate social hierarchies.

Based on this realization, Pasquinelli observes the growing alienation of workers in the face of AI systems that advance by replacing both physical and cognitive labor, as well as the management and control over work. Machines and algorithms increasingly assume the role of supervising and organizing labor, diminishing workers' autonomy and multiplying precarious forms of labor, as observed in gig economy platforms. AI systems not only automate tasks but also impose knowledge hierarchies and reinforce inequalities, continuing the historical process of alienation initiated by the introduction of machinery in the industrial age.

In general, the introduction of AI in the work process further dehumanizes the workers, turning them into mere performers of tasks monitored and organized by automated systems, increasing the gap between control and the execution of labor. This alienating process is intensified by the concentration of power in the large companies that control AI infrastructure, placing workers in a position of greater vulnerability and subordination to these systems. The technical autonomy of intelligent machines is not truly autonomy in the human sense but rather an expansion of machines' capacity to follow instructions or adapt to patterns present in data. These findings allow us to say that actually existing AI serves the reproduction of a highly concentrated economic system that is socially asymmetric.

It is unlikely that we will be able to predict all the effects of automated systems on society. However, drawing on Winner's work, we can assert that even the most commonplace everyday practices are being transferred to these systems and their distant owners, who now govern contemporary life. The technological imperative is confirmed as we organize our lives around sociotechnical apparatuses we do not control, which dictate the pace of culture, relationships, interactions, and fundamental social organization. In this scenario, the loss of individual agency is so profound that even decisions about familiar routes through cities are no longer made without consulting mobile applications.

Even writing itself is being shaped by algorithmic systems, with summaries generated by artificial neural networks and textual interpretations carried out according to the statistical parameters of generative AI models. Consequently, formal education is directly affected by solutions that promise to accelerate learning and reduce the learner's effort, which appears to herald a tragedy foretold. Winner demonstrated that modern technical systems, beyond producing unforeseen effects, can become ungovernable. Initial intentions are insufficient to prevent the resulting disorder and its side effects. A series of enforced changes are driven by the internal logic of artifact developers and technology corporations, who claim to act on behalf of a desire for expanded convenience, supposedly embodying the ethos of humanity.

7.5 Final Considerations

By promoting the centralization of power and knowledge, the actually existing Artificial Intelligence systems reinforce a process of technical alienation that deepens inequalities and the concentration of capital. This phenomenon, when understood through the lens of Langdon Winner (1978, 1986), reveals that the apparent autonomy of technologies is not about true machine independence but rather a dynamic that amplifies control by large corporations and technological elites. The concentration of infrastructures, data, and algorithms prevents society at large, including workers and specialists, from having control or even access to the decision-making processes behind these technologies. Ultimately, what seems to be a democratizing technological revolution hides an authoritarian megatechnology that increases labor precariousness and diminishes human autonomy.

Despite having a different perspective, the application of the concept of alienation that Winner makes can align with Gilbert Simondon's (1980) critique of the separation between culture and technology. For Simondon, the distance between humans and technical objects leads to a lack of understanding of technicality, a phenomenon that, through the prism of AI, is exacerbated by corporate structures that hold a monopoly over technological development. This distance prevents individuals and society as a whole from understanding the cultural, social, and political implications of the technologies that shape their lives. Thus, actually existing AI not only centralizes power but also reinforces a cycle of technical alienation in which technologies are seen as neutral entities, dissociated from their political and economic impacts.

Therefore, Artificial Intelligence is embedded in a technopolitical context that reinforces social and economic hierarchies, acting as a tool for concentrating power in the hands of a few. As Winner highlighted, society finds itself facing technological systems whose actions are, for the most part, invisible and uncontrollable to most people. The perpetuation of this technological model increases the dependence on centralized systems and the subordination of workers to these machines, further distancing the possibility of democratic control over technologies. When analyzing this issue, it becomes clear that actually existing AI not only reproduces existing inequalities but also reconfigures society to maintain and expand these power structures.

References

Barthélémy, J. H. (2012). Fifty key terms in the works of Gilbert Simondon. In A. De Boever, A. Murray, J. Roffe, & A. Woodward (Eds.), *Gilbert Simondon: Being and technology* (pp. 203–231). Edinburgh University Press.

Cho, R. (2023, June 9). AI's growing carbon footprint. *State of the planet—News from the Columbia Climate School*. https://news.climate.columbia.edu/2023/06/09/ais-growing-carbon-footprint/

Dyer-Witheford, N., Kjøsen, A. M., & Steinhoff, J. (2019). *Inhuman power: Artificial intelligence and the future of capitalism*. Pluto Press.

Goldman Sachs. (2024, April 28). AI/data centers' global power surge and the sustainability impact. *Equity Research*. https://www.goldmansachs.com/images/migrated/insights/pages/gs-research/gs-sustain-generational-growth-ai-data-centers-global-power-surge-and-the-sustainability-impact/sustain-data-center-redaction.pdf

Hui, Y. (2020). Algorithmic catastrophe—The revenge of contingency. *Parrhesia: A Journal of Critical Philosophy, 34*.

International Energy Agency (IEA). (2024). *Electricity 2024: Analysis and forecast to 2026*. International Energy Agency.

Marx, K. (1971). *O Capital* (Book 1, Vol. 1). Civilização Brasileira.

Maslej, N., Fattorini, L., Perrault, R., Parli, V., Reuel, A., Brynjolfsson, E., Etchemendy, J., Ligett, K., Lyons, T., Manyika, J., Niebles, J. C., Shoham, Y., Wald, R., & Clark, J. (2024). *The AI index 2024 annual report*. AI Index Steering Committee, Institute for Human-Centered AI, Stanford University.

Morozov, E. (2018). *Big tech*. Ubu Editora.

O'Flaherty, K. (2024, August 7). CrowdStrike reveals what happened, why, and what's changed. *Forbes*. https://www.forbes.com/sites/kateoflahertyuk/2024/08/07/crowdstrike-reveals-what-happened-why-and-whats-changed/

Pasquinelli, M. (2023). *The eye of the master: A social history of artificial intelligence*. Verso Books.

Rogoway, M. (2023, February 22). Google's water use is soaring in The Dalles, records show, with two more data centers to come. *The Oregon Tech*. https://www.oregonlive.com/silicon--forest/2022/12/googles-water-use-is-soaring-in-the-dalles-records-show-with-two-more-data--centers-to-come.html

Simondon, G., Mellamphy, N., & Hart, J. (1980). *On the mode of existence of technical objects*. University of Western Ontario.

Winner, L. (1978). *Autonomous technology: Technics-out-of-control as a theme in political thought*. MIT Press.

Winner, L. (1986). *The whale and the reactor: A search for limits in an age of high technology*. University of Chicago Press.

Chapter 8
Are We Still Sleepwalking? Rethinking Technological Somnambulism in the Age of AI

Hao Wang

Abstract Langdon Winner uses the concept of "technological somnambulism" to criticize how we often accept technology-driven changes without fully questioning their deeper impacts. In recent years, there has been a strong push to develop responsible technologies, especially in the age of Artificial Intelligence (AI), that anticipate socio-ethical concerns and integrate human values into technological design. At first glance, this suggests that we are no longer "sleepwalking" through technological change but are instead taking deliberate steps to ensure that AI aligns with social values. However, this paper argues that Winner's warning about technological somnambulism remains relevant today. Drawing on a multi-level framework of AI ethics, the paper shows how current efforts toward responsible AI often focus narrowly on anticipating and addressing impacts at the artifact level, while failing to challenge the deeper structural forces that perpetuate power imbalances and further marginalize vulnerable groups. This neglect of structural impacts risks leading us into what I call "AI somnambulism". To move beyond this, the paper calls for strengthening existing ethical frameworks with a "structural imagination" to shift our focus from specific artifacts to deeper, structural issues in AI. It concludes by outlining how such imagination can be translated into more practical design practices, aiming to foster a truly responsible AI future.

Keywords Langdon Winner · Autonomous technology · Responsible AI · Structural issues · AI ethics

---

H. Wang (✉)
Department of Social Sciences, Wageningen University & Research, Wijchen, Gelderland, The Netherlands
e-mail: hao2.wang@wur.nl

© The Author(s), under exclusive license to Springer Nature Switzerland AG 2026

A. Fox, D. Cressman (eds.), *Arrangements of Power: Tracing Langdon Winner's Legacy Within and Beyond the Philosophy of Technology*, Philosophy of Engineering and Technology 50, https://doi.org/10.1007/978-3-032-17371-3_8

8.1 Introduction

Langdon Winner (1986) uses the metaphor of "technological somnambulism" to describe how society often adopts new technologies passively, as if they are "sleep-walking" into their broader impacts, rather than consciously deciding how these technologies should shape their lives or align with their values. Unlike technological determinists, he believes that technological progress is shaped by ongoing choices and decisions made by humans and society. However, we often move through these decision points in a kind of sleepwalking state, unaware of the deeper implications that come with embracing new technologies. This "sleepwalking" state makes technological advancement seem inevitable, as if it is "out of control" or becoming autonomous (1978).

It is acknowledged that over the past two decades, there has been a strong push to develop responsible technology that anticipates the socio-ethical implications of design and integrates human values into tech. Numerous regulations have been introduced to guide and govern technological development, ensuring that innovation occurs in a way that accounts for societal values and ethical concerns (Jobin et al., 2019). Moreover, approaches such as Value-Sensitive Design (VSD) and other "by design" frameworks explicitly integrate politics, ethics, and social dimensions into the development and implementation of technology (Brey & Dainow, 2023; Dignum et al., 2018; Friedman et al., 2013). These frameworks emphasize proactively embedding values like privacy, fairness, and transparency into technological systems, highlighting a shift away from the passive acceptance of technological impacts. These growing efforts seem to suggest that we are no longer sleepwalking with technology but are instead taking deliberate steps to critically engage with its implications and ensure it aligns with human values.

However, this paper shows why Winner's account of technological somnambulism is still relevant today and how the idea can be translated into more practical, responsible design practices. I want to argue that although those growing efforts of regulation and by-design approaches are highly important to make technology responsible, they tend to only focus on addressing artifact-level issues that can be fixed by technological design alone. This focus, however, makes them often overlook the deeper, structural patterns that shape our society in more profound ways (Wang & Blok, 2025; Ryan et al., 2024; Wang & Blok, 2026). This narrow focus is exactly what Winner criticizes as "sleepwalking" into the deeper impacts of technologies. Thus, I reinterpret Winner's technological somnambulism within this structural context, as a critique of the way society sleepwalks into the structural impacts of technologies. These structural impacts differ from the effects at the artifact level, which are typically caused by individual technologies or specific groups and often involve unintended consequences like data bias or algorithmic transparency issues. Structural effects, on the other hand, emerge from broader socio--political forces (such as systemic inequality and power dynamics) and ontological changes (like disruptions to everyday life). These impacts are more deeply embedded in socio-political systems and ontological structures.

To illustrate this, I examine the discourse of developing responsible Artificial Intelligence (AI) in our society. I will show that despite significant efforts to develop responsible and trustworthy AI, we might still be sleepwalking into AI's widespread, structural impacts. This is because we often focus on the immediate, artifact--level issues, such as data bias or algorithm transparency, while ignoring the deeper, structural effects. These structural impacts require more collaboration, work, and systemic transformation, but are often overlooked or dismissed in practical technological design. As a result, they become the "elephant in the room," ignored despite their importance. I have chosen responsible AI discourse not just because AI is becoming a symbol of our technological society, but AI may be the most suitable in describing the topic due to its wide and deep influence on humans and society.

This paper is structured into four main parts. First, I will revisit Winner's idea of technological somnambulism and suggest why it should be interpreted as focusing on the structural impacts of technology. Next, I develop a multi-level framework to further explain what the structural-level issues are by drawing from philosophical, social, and political theories. Third, I will apply this framework to show why current AI ethics remains unaware of deeper structural issues, leading to a kind of "AI sleepwalking". Last, I will propose the concept of "structural imagination" as a way to help address AI sleepwalking and consider how to apply it to in responsible AI design.

8.2 Unpacking the Idea of Technological Somnambulism

In *The Whale and the Reactor* (1986, 5–10), Winner introduces the concept of "technological somnambulism" to describe the phenomenon that "we so willingly sleepwalk through the process of reconstituting the conditions of human existence" when developing and using technologies (10). He argues that while people actively design, create, and use technology, appearing rational and deliberate in their decisions, they are, in reality, in a state of low awareness, failing to recognize how these technologies profoundly reshape human existence. Sleepwalking, a state between wakefulness and unconsciousness, serves as a metaphor for this paradox: we act as if we are in control, yet we rarely care about the broader and deeper influence technology brings. In this sense, Winner's idea of technological somnambulism is not just a critique of society's neglect of these broader impacts but highlights how this passive stance creates an illusion of awareness and control, when in fact, most people remain unaware of technology's far-reaching impacts.

Winner intentionally uses the term "technological somnambulism" to, on the one hand, recognize that technology may have a tendency to get out of human control and, on the other hand, emphasize that he rejects technological determinism or essentialism (Winner, 1986). This term of somnambulism suggests that humans can still wake up and that technology only appears to get out of control or become autonomous because we humans are either unaware of or unwilling to take an active role in steering it. Winner shows that he deliberately uses the term "technological somnambulism" to replace the idea that "technological innovation is the basic cause of changes in society, and humans have no choice but to watch this inevitable process unfold." (1986, 10) As he argues, technological determinism is "much too strong," and sleepwalking is a more fitting metaphor because humans and society fundamentally have the capacity to regain control over technological forces, shifting the course of technological transformation in society (9–10).

To fully understand this "sleepwalking" phenomenon, we need to understand what Winner calls "a more subtle version of autonomous technology," which he presents as an alternative to technological determinism (Winner, 1993, 371). This helps clarify what exactly Winner is asking us to recognize about the deeper effects of technology and why we tend to ignore them. In *Autonomous Technology* (1978), Winner describes autonomous technology as the idea that "somehow technology has gotten out of control and follows its own course, independent of human direction" (13). He draws on the work of thinkers like Bruno Bettelheim, John Kenneth Galbraith, and, most notably, Jacques Ellul. According to Ellul, "Technique has become autonomous; it has fashioned an omnivorous world which obeys its own laws and which has renounced all tradition" (Ellul, 1964, 14). In other words, technology develops according to its own internal logic, seemingly with a life of its own (ibid. 159).

When reading about technological advancement becoming autonomous, many scholars may think that Winner has simply overlooked something obvious. For them, it is clear that technological development is never an autonomous process that follows its own path, but rather that all technologies are socially created and chosen, reflecting humans' needs, ideas, values, and goals. For example, the development of a smartphone is not the result of an autonomous path of technical perfection, but rather a social process involving various groups: Tech companies, consumer preferences, government regulations, labor conditions, and global supply chains all play a role. Therefore, speaking of technology's autonomy might seem misguided, as all aspects of technical systems emerge from negotiations among social actors with different levels of power. This leads to the common denial of the idea of autonomous technology, with critics calling for moving beyond the determinism of certain thinkers (Bucciarelli, 1994).

This common rejection of autonomous technology might come from not fully grasping what technological autonomy really means for Winner. If we take a closer look, it is clear that Winner is fully aware the active role that humans and society play in developing and using technology. His point is not that technology emerges outside of social influence, but that, in modern society, these social groups are so deeply embedded in the technological environment that their actions often revolve around the relentless pursuit of efficiency and technological logic (Winner, 1995). Even when different groups have varying degrees of power, their decisions tend to be driven more by the push for productivity and technological means, rather than by independent or free choices. That is why Winner thinks that Ellul's use of the term "Technique" or "la technique" is more appropriate over "technology" (Winner, 1978). For Ellul, technique is not simply about specific tools or machines; it is about a total system of methods designed to maximize efficiency across all human activities, from work and economics to leisure and even personal relationships (Ellul, 1964).

Therefore, Winner's idea of autonomous technology or "out of control" refers to a deeper, structural-level phenomenon. This phenomenon concerns the overall tendency of technologization in our society, where every domain, from governance to daily life, becomes increasingly dominated by technological logic. This phenomenon is not about the use of specific technologies but rather the deeper *pattern* of interconnected and interdependent technologies influencing society beyond individual decisions or policies, often in ways that are difficult to predict or control. For example, "technological drift" is the idea that technology tends to develop and advance in ways that are not always planned or intentionally guided by human actors, creating unforeseen consequences (Winner, 1978). Another is the "technological imperative", which suggests that once a technology becomes feasible, there is an almost inevitable drive to develop and implement it, regardless of whether it is necessary or beneficial (1978). Winner also critiques what he calls "the cult of innovation" (Winner, 2018), the widespread belief that technological progress is inherently positive, inevitable, and always leads to improvements in society. All these phenomena Winner describes are not about any particular technologies but reflect a general trend of unchecked technological advancement, the tendency to prioritize technological logic over ethical or social considerations.

However, these big-picture impacts of technological development are increasingly disregarded since the 1980s, largely due to the rise of social constructivism and the empirical turn in the philosophy of technology (Winner, 1993). These new approaches shifted the focus away from general theories about technology and toward the empirical study of specific artifacts and engineering practices (Achterhuis, 2001; Kroes & Meijers, 2001). Scholars became less interested in discussing technology as a broad, overarching phenomenon ("Technology with a big T") and instead concentrated on concrete aspects, such as the design and implementation of specific technological systems. Winner criticizes this shift in his paper "*Upon Opening the Black Box and Finding It Empty*":

> They usually find it sufficient to gather evidence of social activities most clearly connected to technological change. Insofar as there exist deeper cultural, intellectual, or economic origins of social choices about technology or deeper issues surrounding these choices, the social constructivists choose not to reveal them. (1993, 371)

As a result, this empirical turn has led to a strong emphasis on examining specific technological artifacts, such as smartphones, self-driving cars, or data management systems, while paying less attention to their deeper, structural effects on society.

Building on this analysis, we can interpret Winner's idea of technological somnambulism by making his structural considerations more explicit: *while humans may appear deliberate at the artifact level, they are actually sleepwalking by ignoring the deeper, structural impacts of technology*. Winner does not explicitly use the term "structural issues" to describe all the deep impacts of technology. However, in his books and papers, he often uses terms like "deeper patterns" and "structures" to highlight influences that go beyond the narrow focus on individual technologies

(Winner, 1986, 1995). For example, Winner criticizes social constructivists' response to the idea of technological autonomy as "almost always superficial, dealing with surfaces rather than deeper patterns." (1995, 67) Also, he explicitly highlights the importance of considering "structures" in technological development, linking it to some "basic conditions" underlying social practices of making technology, such as social classes (1993). At times, he frames "structures" in a cultural sense, referring to a "basic metaphysical disposition" or the broader ideology of technological "rationality" (1993, pp. 370–371). Despite these different insights, Winner believes that if we focus only on artifact-level issues while ignoring the deeper, structural impacts of technology, we risk sleepwalking into a world where technology slips beyond human control (1978). In this sense, Winner's technological somnambulism does not mean people ignore technological impacts altogether, but rather that they focus only on the immediate effects at the artifact level while overlooking the more structural forces at play.[1]

To sum up, Winner's use of technological somnambulism criticizes how technological impacts are often treated as isolated issues related to individual artifacts, rather than being understood within the broader structures that shape them. But while he talks about things like the technological myth and the technological imperative, he does not provide a systemic analysis of the deep, structural impacts that people are sleepwalking into. To really take Winner's critique and turn it into something actionable for designing responsible tech, we need to get more specific about what those structural impacts are. For the rest of the paper, I will (1) develop a theoretical framework for understanding the structural impacts of technology, (2) apply this framework to AI context to see how the structural impacts are overlooked in AI ethics discussions, and (3) explore how to translate structural considerations into responsible AI design practices.

8.3 Understanding Multi-level Impacts of Technology

This section distinguishes between two different levels of technological impact: the artifact level and the structural level. Winner mentions a similar way of seeing technological influence on different levels. For example, he holds that "socially constructing technologies succeed at some level of practice, they sometimes undermine what are or ought to be key concerns at another level." (1995, p. 371) As he believes, technologies are socially constructed at the artifact level, but we have to acknowledge the possibility that "the ebb and flow of social interaction among social groups may reflect other, more deeply seated processes in society" (371). However, Winner

[1] To be more explicit and for convenience, I use "structural issues" to refer to Winner's meaning of deep and systemic effects of technologies. "Structural" and "systemic" can have subtle differences, as distinguished in Sally Haslanger's paper (2023). However, they are generally used interchangeably, as they are in this paper. In a later section, I will provide a more detailed analysis of what "structural" means and what makes an issue structural.

himself does not provide a systematic analysis of the two-level framework. To complicate things further, for structural issues themselves, it is still unclear what exactly a "structure" is and what makes an issue "structural". This section will develop a multi-level framework to analyze these technology's different impacts, drawing on theories from philosophy, social, and political thought.

Let's start with technological impacts at the artifact-level. In this paper, if an issue/impact is mainly caused by how a specific technology is developed and used, we call it an artifact-level issue. At this level, each of these technologies is developed within its own specific context, designed to perform a particular function. For example, a particular brand of refrigerator raises concerns about energy consumption and environmental sustainability. A certain type of car could contribute to pollution and traffic safety issues, while some specific medical devices such as pacemakers or insulin pumps must ensure reliability and accessibility. Common ethical and societal issues at this level include product safety, environmental impact, accessibility, and responsible resource use, etc. To develop these technologies responsibly, manufacturers and designers must consider these issues from the start, improving safety standards, reducing environmental impact, and ensuring fair access. Addressing these concerns at the artifact level could largely make individual technologies more ethical and socially responsible.

When it comes to structural impacts, we need to explain them in more detail. Not only are these structural issues central to Winner's concept of technological somnambulism and his philosophy of technology, but these issues themselves are not immediately clear.

What does a "structure" mean? In social and political philosophy, as Sally Haslanger (2023) explains, a "structure" or "system" refers to something that is greater than the sum of its parts. For example, a forest is not just a collection of trees, and a band is more than its members (it also includes roles, shared beliefs, and values that define the group). A structure can be seen as an abstract framework of these interconnected patterns, or a "schema" as Sewell puts it. However, both Iris Young (2001) and Haslanger (2023) argue that structures are not merely abstract; they actively shape the realities people experience. These structures influence the opportunities and life chances of individuals within them. In this sense, structures are a combination of patterns and actions that go beyond abstract relationships and exert a real-world impact, determining the distribution of resources and shaping lived experiences (Wang & Blok, 2025).

Young's rich study on structural injustice also offers a helpful way to think about what it is like to be a structural issue. For her, a "structural issue" is one that is shaped by larger structures, not just any individual actions. These structures often exist long before and after any one person's involvement, so individuals are not necessarily the creators of them. For example, sweatshop labor is tied to the larger structure of global capitalism. Workers in developing countries are stuck in a certain class position compared to factory owners, who rely on contracts from multinational companies. These companies compete in a global market, and consumers keep the system going by constantly buying new clothes, influenced by the fashion industry's push for the latest trends. This desire for "newness" becomes a part of what Young (2011) calls "habitus", namely, social habits and norms that make it hard to break free from the cycle. Even if individuals make more ethical choices, like buying fair-trade or second-hand clothing, those actions alone may not immediately fix the deeper, structural forces at play.

So, we can define structural issues as the concerns that are created and maintained by large-scale systems and patterns that are beyond the control of any single person. They are built into the way society works, so they keep going, even when individuals try to make different choices. These problems persist because they are deeply woven into the systems that shape our lives. That said, individuals still play a role in maintaining these structures. Drawing on Giddens's "structuration" theory, Young explains that when people act, they are both pursuing their own goals and, in the process, reinforcing the structures that guide their actions (Young, 2011). Take sweatshop labor, for instance: it exists independently of any one person's choice, but when consumers buy products from companies that rely on such labor, they unintentionally support and continue these exploitative practices. This dynamic shows how structures shape individual actions, which in turn help reproduce those very structures.

Structural influences can vary in complexity and scope. As mentioned earlier, Winner's concept of "structure" is sometimes linked to socio-political structures (e.g., social classes), while at other times, it is framed in a more cultural or philosophical phenomenon (e.g. technological rationality). To clarify, I thereby further differentiate two types of structural issues: those related to socio-political structures and those tied to ontological structures.

The socio-political structures are rooted in existing unequal social and political structures, which involve what Winner calls "basic condition" or what Young calls the institutional background and processes that "fundamentally condition the opportunities and life prospects of the persons located in those positions" (Young, 2001, p. 14). In this sense, these socio-political structures exert a type of structural force that does not come from direct interactions between individuals but operates on a larger scale. This power is often seen through established norms, policies, and hierarchies deeply ingrained in our social, political, economic, and cultural systems. A classic example is Winner's description of the Long Island Bridge, which was deliberately designed with low-clearance arches that prevented buses from passing through (Winner, 1986, p. 22). This design effectively restricted access to the affluent areas of Long Island for lower-class communities, particularly those reliant on public transportation. The bridge's design reflects how technology and infrastructure can reinforce existing socially unequal power structures and maintain existing class divisions. These socio-political structures are more stable: While individual artifacts may not have a lasting impact, societal-political structures, such as social classes or racist systems, have persisted over time, continuously shaping individual artifacts.

Technological impacts at the ontological level refer to deeper philosophical and existential concerns about how technology changes who we are, how we see ourselves, and how we act in the world. For example, the invention of the train in the nineteenth century did more than simply introduce a new mode of transportation; it reshaped how people experienced time and space (Blok, 2025). Trains compressed distances and allowed for faster travel, fundamentally changing how individuals perceived their connections to distant places and people. This new infrastructure changed how people experienced time. Trips that used to take days or weeks could now be done in just a few hours. The widespread use of trains not only redefined the physical landscape but influenced societal structures, creating a new sense of immediacy and connectivity that transformed social and economic interactions. Also, this invention of trains may reflect the wider phenomenon of "cult of innovation," a cultural obsession with innovation that glorifies new technologies as inherently good and always leading to progress (Winner, 2018, p. 71). These ontological structures (e.g. our beliefs, mindsets, and worldviews) are more enduring and stable, representing deeper, more foundational processes that are even more "structural" than socio-political structures.

In summary, we can identify technological impacts at the artifact, socio-political, and ontological levels (see Fig. 8.1). Following this multi-level framework, we can see that Winner's concept of technological somnambulism critiques the focus on individual artifact-related issues, while overlooking the technological impacts at both the socio-political and ontological levels. These structural impacts involve existing inequalities in social and political structures, as well as how technology fundamentally alters how people view the world, shaping their behaviors and interactions with others. In the next section, I will apply this framework to the responsible AI context to explore the specific structural impacts that have been overlooked.

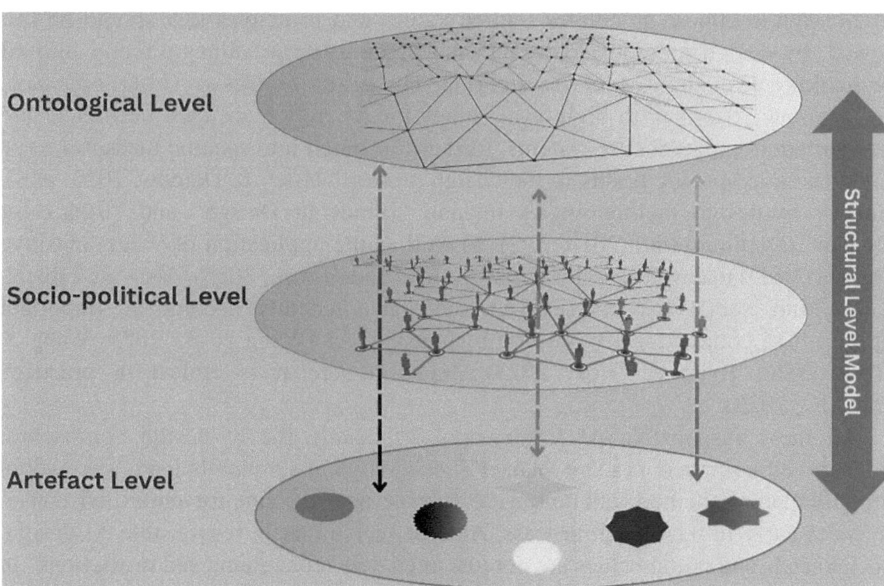

Fig. 8.1 The multi-level impacts of technology on humans and society

8.4 AI Somnambulism

This section will illustrate technological somnambulism specifically in the context of AI. Building on Winner's idea, I use the term "AI somnambulism" to describe how we accept AI-driven changes or develop responsible AI without fully questioning their deeper, structural implications. I will explain why we tend to sleepwalk into AI's structural impacts and then use the multi-level framework to explore what the structural issues that are often missing.

At first glance, AI ethics and governance might give the impression that we are fully aware of AI's impacts and are actively regulating it, aiming to steer it towards the ideal of "maximizing the benefits of AI systems while minimizing their risks" (AI HLEG, 2019, p. 4). We have a guidelines-based approach, with top-down frameworks published by governing bodies, NGOs, or major tech companies to help map out what important values and principles in making AI responsible. For instance, Corrêa et al. (2023) compiled around 200 guidelines, including well-known examples like the OECD AI Principles (OECD, 2019) and the EU's Trustworthy AI guidelines. While these guidelines' general and abstract nature often make them difficult to apply to specific AI design, they are useful for identifying potential values and principles for developing responsible AI, such as transparency, justice and fairness, non-maleficence, responsibility, and privacy (Jobin et al., 2019).

We also have more practical approaches, like various by-design approaches. They are known for integrating ethical values into AI development by translating abstract principles into concrete design practices (Dignum et al., 2018; Sadek et al., 2023). While ethical guidelines outline values and principles that should be followed, by-design methods focus on incorporating these values directly into AI technology throughout its entire lifecycle. The Ethics by Design (EbD) approach, for example, "starts with high-level values for AI, which are then translated into design requirements for AI systems, further translated into specific measures to be undertaken at specific points in the design process" (Brey & Dainow, 2023, p. 3). Similar by-design methodologies include "Ethics in Design" and "Ethics for Design" (Dignum et al., 2018, p. 2), as well as the application of Value-Sensitive Design (VSD) to develop 'value-sensitive AI' (Sadek et al., 2023). Also, the Ethical, Legal, and Social Aspects (ELSA) Labs have recently become an important approach to collaboratively develop responsible AI (Wang et al., 2025; Wang & Blok, 2026; Ryan & Blok, 2023), "making sure it is ethical in practice" (NLAIC, 2023).

All these responsible AI frameworks, especially the by-design approaches, appear well-equipped to solve Winner's technological somnambulism. They adopt an anticipatory method, attempting to foresee and account for undesired consequences early in the design process. Also, it seems that in responsible AI design, social needs and moral values are not just treated as "constraints and restrictions" in AI development that need to be managed to serve its economic and technical goals (Von Schomberg, 2013, 16). Rather, the integration of values signifies that we have put ethical considerations at the center of AI development, seeing ethical and social values as a driving force and a core operational principle to AI innovations. In this sense, AI ethics has moved beyond Winner's critique of technological somnambulism by ensuring that AI is designed and regulated in a value-based way, keeping it firmly under human control and aligned with human values.

In this paper, I acknowledge that AI guidelines play a crucial role in identifying key values and principles, while by-design approaches like VSD, EbD, and ELSA Labs are essential for integrating those values into AI artifacts. However, I argue that these approaches are not sufficient because they often focus on designing responsible AI at the artifact level, missing deeper structural issues that are usually too abstract and complex to be solved through technical design alone. In this sense, we might be "sleepwalking" when relying on these methods for responsible AI development. In the rest of this section, I will analyze AI's impacts at the artifact, socio-political, and ontological levels to see what artifact issues these approaches focus on and what structural impacts they tend to ignore.

8.4.1 AI Impacts at the Artifact Level

At the artifact level, socio-ethical concerns emerge in direct relation to the development and use of specific AI technologies. For example, facial recognition systems used in law enforcement may raise concerns about privacy and racial profiling (Eubanks, 2018). Autonomous vehicles may create safety issues (Goodall, 2014). Credit scoring systems could be problematic if their decision-making algorithms are not transparent enough (Sax & Wang, 2025; Wang, 2022b). Other typical socio-ethical issues at this level include algorithmic bias (Franke, 2022), algorithmic fairness (Wachter et al., 2021), and accountability (Novelli et al., 2023), among others. Addressing these issues can significantly improve the responsibility of the specific AI artifact.

AI guidelines and by-design ethics approaches tend to focus on mitigating these artifact-level issues. This is directly linked to their pragmatic approach to values. As shown earlier, by-design approaches often start with broad AI values, which are then translated into a series of design requirements for AI technologies and further broken down into specific actions throughout the design process. So, they are fundamentally driven by a sort of pragmatism, the idea of integrating only those values that can be operationalized within AI systems (Blok, 2025). To achieve that, they often prioritize the values that can be more easily implemented in concrete design practices. The goal of "trustworthy AI," for example, often prioritizes a handful of technically-fixable values such as privacy, transparency, security, and accountability (AI HLEG, 2019). While these are important, they are also the easiest to formalize mathematically and implement as technical solutions (Hagendorff, 2020).

If some values carry deeper political and historical weight, like fairness and justice, by-design approaches often focus only on their technically fixable aspects. For example, they address fairness and justice by targeting data and algorithmic issues, such as reducing data bias or improving data diversity (Benjamin, 2019). This allows values to be successfully translated into concrete, actionable design tasks that engineers can follow to fix. However, fairness and justice are actually deeply tied to complex historical and social structures, such as racism, sexism, and post-colonial legacies, that shape biased data in the first place (Madaio et al., 2022). Moreover, deciding who gets to define "fairness" are inherently political choices. By-design approaches often ignore these deeper social and political dimensions, which cannot be solved by design alone but require broader discussions, negotiations, and societal engagement.

Thus, while AI guidelines and by-design ethics help address artifact-level concerns in developing responsible AI, they are not sufficient. AI is deeply embedded in socio-political and ontological structures, meaning socio-ethical concerns cannot be fully addressed by focusing only on individual technologies. To grasp AI's full impact, we must go beyond the artifact level and examine its deeper, structural influence.

8.4.2 AI Impacts at the Socio-political Level

AI systems are embedded in power dynamics, social norms, economic disparities, and systemic inequalities. This means they can reinforce, maintain, or challenge structures like colonialism, sexism, racism, and class divisions. For example, surveillance capitalism (Zuboff, 2019) exploits personal data, AI access remains unequal (Muldoon & Wu, 2023), and corporate control over AI limits democratic oversight (Hoeksema, 2023). AI also contributes to job displacement and labor exploitation (Gray & Suri, 2019) while reinforcing Western dominance over the global South (Birhane, 2020; Couldry & Mejias, 2018). Addressing AI's socio-political impacts requires recognizing these deeper power structures, not just fixing individual AI artifacts. Here, I outline several AI impacts that are closely tied to power dynamics.

When integrating values into AI, by-design approaches often assume some values are universally shared and ready for implementation. However, choosing which values to integrate into AI is inherently political. While AI ethics frameworks often claim to be based on international consensus or human rights principles (AI HLEG, 2019; Brey & Dainow, 2023), it is questionable whether these truly reflect global perspectives. For example, autonomy is frequently presented as a universal value, but it is deeply rooted in Western individualism and may hold different meanings in other cultures (Nakada & Tamura, 2005). Also, current AI ethics frameworks are largely shaped by Western institutions and corporations, with limited input from the global South (Birhane, 2020). This raises concerns about the universality of these values in the global AI discourse.

Public participation is often praised as a key component of responsible AI design, yet it is often influenced by power dynamics. Sherry Arnstein (2015) warns the possibility of "manipulation", a way to create the illusion of participation without actually giving people a say. When power imbalances exist, participation can feel more like a checkbox exercise than real involvement (Gilman, 2022). In AI development, this happens when those with the most influence decide how participation works (Blok & Lemmens, 2015). Marginalized groups might be included, but only as data points or to make AI projects seem ethical, while their actual concerns remain unaddressed. This may even lead to a vicious cycle of participation, when people see that their input does not change anything, they lose trust and interest in the process, making future participation even harder (Wang, 2022b). To make AI truly inclusive, we need to challenge these power imbalances.

AI technologies also tend to benefit those with power and resources while further excluding marginalized groups. Many AI systems are designed to serve the interests of dominant economic and social groups, reinforcing existing inequalities (Ryan et al., 2024). For instance, expensive farming robots are often accessible only to large landowners, leaving smallholder farmers behind. Similarly, AI-driven hiring algorithms may favor candidates from privileged backgrounds, further marginalizing underrepresented communities. This exclusion is not always intentional but results from systemic biases in data, design choices, and access to technology. Since AI is mostly developed in wealthy countries and for mainstream groups, many systems fail to work effectively in diverse contexts. If we do not actively address these structural inequalities, AI could end up deepening social divides instead of helping to bridge them.

8.4.3 AI Impacts at the Ontological Level

At the ontological level, AI is not just about individual artifacts but embedded in and reinforcing a deeper structure of *datafication*. This process reshapes human existence by turning everyday life into quantifiable data used for decision-making, prediction, and optimization. This process alters the way we understand and experience the world. For example, when AI technologies are applied in smart cities, they collect data through sensors to generate specific artifacts, but at the same time they reinforce a broader ontological structure of datafication, where interconnected technologies create a digital ecosystem that influences human actions (Blok, 2023). This datafication process not only establishes a new infrastructure for daily life but reshapes our fundamental assumptions that data is the foundational element of reality. These shifts influence how we engage with the world, raising ontological challenges in the digital age, such as the detachment from real world, the disruption of traditions, and the rise of dataism and AI solutionism, pan-computationalism, AI alienation, pan-surveillance (Zuboff, 2019), dehumanization (Fritts & Cabrera, 2021), and objectification (Wang, 2022a, 2023).

For example, smallholder farmers often feel overwhelmed by the adoption of AI (van der Burg et al., 2022). They have to choose between adopting it or risking falling behind. This is part of a bigger issue: datafication disrupts traditional farming. It is not just about turning things digital; it changes how we think about everyday tasks

(Blok, 2023). Datafication assumes turning everything into data leads to better efficiency and insights (Korenhof et al., 2021), but smallholder farmers might care more about community and eco-friendly practices than just being more efficient. So, the structural impacts of datafication could disrupt traditional practices and foster AI solutionism.

AI also challenges traditional notions of agency and autonomy. As AI systems make decisions, generate content, and even simulate human-like reasoning, like ChatGPT and DeepSeek, they blur the lines between human and machine intelligence. This raises questions about what it means to think, create, or make choices in an era where AI plays an active role in shaping information, culture, and even our personal interactions.

Moreover, the widespread adoption of AI contributes to a growing sense of AI inevitability, a belief that tech is just moving forward whether we like it or not, and all we can do is adapt. Think about ChatGPT: no one really knew about it until it appeared, and suddenly it reshaped almost everything from work and education to communication and everyday lives, forcing people to adjust, whether they were ready or not. This feeling makes AI seem unstoppable, and it can stop us from questioning whether AI is really benefiting society or just serving corporate and economic interests. And the bigger question: do we really even need AI in the first place?

In sum, this section explains the risk of AI sleepwalking when developing responsible AI. It shows how responsible AI practices often focus on individual

Table 8.1 The multi-level impacts of AI

Responsible AI	Focus	Typical issues
Artifact Level	The design and use of specific AI artifacts	• Data bias • Privacy • Algorithmic fairness • Explainability/ Transparency, etc.
Socio-political Level	AI is embedded in power dynamics, interacting with historical and social structures	• Surveillance Capitalism • Data Colonialism/ Imperialism • Data racism • Inequitable access to AI technologies, etc.
Ontological Level	Overall structure of datafication conditions humans' understanding and experiencing the world	• Dataism/ Pan-computationalism • Objectification of human/ non-human beings • Detachment • AI solutionism • Disruptions of traditions/ meaning, etc.

artifact-level issues, overlooking the broader structural impacts they may cause. These different levels of AI impacts can be briefly summarized in Table 8.1.

8.5 Awakening: Towards a Structural Imagination

This section explores how to integrate structural considerations into practical AI design to help address the issue of "AI sleepwalking". As mentioned earlier, Winner's concept of sleepwalking is not rooted in technological determinism—he argues that humans still have an active role in shaping technology's advancement. In some of his work, he does mention the role of citizens in design, such as through participatory design (Winner, 1992), but he does not provide a clear framework for how this would work in practice. This gap makes it difficult to translate his critical perspective into concrete strategies for responsible AI design.

By rethinking Winner's technological somnambulism, I propose that in responsible AI design, we need to develop a *structural imagination*, an ability to connect challenges in individual AI artifacts to broader social issues and ontological structures. This notion of structural imagination is inspired by Wright Mills' (2000) concept of "sociological imagination," which involves recognizing the link between personal challenges and larger social structures. Mills argues that people often lack this structural reflection, only focusing on their "personal troubles" and "immediate relations with others", which prevents them from realizing that these individual problems are connected to larger "public issues of social structure" (ibid. 8). As Mills explained, we need to "consider the economic and political institutions of the society, and not merely the personal situation and character of a scatter of individuals." (ibid. 9) For instance, assigning blame for homelessness solely to individuals living on the streets, based on assumptions about their personal choices, such as laziness or lack of motivation, fails to acknowledge the societal factors contributing to homelessness. This view ignores the role of social challenges like inadequate schools, job shortages, high housing costs, and lack of family support, all of which can lead to homelessness.

In the realm of responsible AI design, we require this kind of imagination to shift our focus from individual AI artifacts to structural issues. It is essential not just to tackle artifact-level issues but to grasp the harm caused by the systemic effects of 'structures' that AI technologies create, maintain, and reinforce. This structural effects compel individuals, regardless of the good intentions behind AI design, to act in ways that can result in significant harm to certain groups of people. In this light, trying to impose what is considered a responsible AI on society without recognizing these systemic forces could lead the society toward irresponsibility. Therefore, this structural imagination serves as a critical and transformative approach to uncover these structural issues, questioning the unspoken assumptions in current AI ethics discourses and exposing blind spots in AI guidelines, policies, and regulatory frameworks.

Table 8.2 The Structural Impact Matrix used to encourage structural imagination in responsible AI design

Structural level	Impact	Issue	Diagnostic question
I. Socio--political	1. Existing Power Structure	• Socioeconomic inequalities • Power asymmetries	*Before we implement this AI system, what existing power imbalances or marginalized groups should we consider in this domain?*
	2. Reproducing Injustice	• Injustice • Unequal distribution of benefits	*Does this AI system serve all communities fairly, or are some groups at risk of being excluded from its benefits?*
	3. Structural Change	• Relational responsibility	*In what ways can this AI system contribute to a more equitable and just society?*
II. Ontological	4. Mindset	• Dataism • AI solutionism • Anthropocentrism	*Are there underlying beliefs or biases influencing how this AI system is built?*
	5. Lifestyle Influence	• Disruption of traditions • Detachment • Datafication • Loss of meaning	*What broader effects might this AI technology have on humans' daily life (e.g. habits, behaviors, or lifestyles), and at what point could those effects become problematic?*
	6. Alternative Solutions	• Design for marginal groups • Community-based design Social innovations	*Are there alternative (non-AI) approaches that might be more effective or ethical in addressing this problem?*

While I cannot go into all the details here, I briefly introduce a structural impact matrix to encourage structural imagination in participatory AI design. This matrix is intended for the early stages of design, when AI systems can still be adapted. It helps AI developers and stakeholders identify potential issues and make responsible changes. The matrix is not meant to be a strict checklist for developers; it is designed for use in workshops or focus groups where stakeholders can have meaningful discussions. As I have highlighted in this paper, many socio-ethical issues, particularly structural ones, cannot be solved by design alone. They may need dialogue and negotiations from diverse stakeholders to find solutions. That is why the questions in the matrix are designed to pinpoint relevant issues at structural levels, while still being open-ended enough to encourage varied perspectives and foster in-depth discussion (Table 8.2).

8.6 Conclusion

This paper rethinks Winner's concept of technological somnambulism and explores its relevance for reflecting on responsible AI practices. We define this somnambulism as a critique of the tendency to focus only on artifact-level impacts of technology, while ignoring the deeper, structural issues that often cannot be addressed by design alone. I further explore some structural issues that are often overlooked in responsible AI design. Finally, I propose the idea of "structural imagination" to counter against AI sleepwalking and offer practical ways to implement it in real-world design. As Winner shows, addressing these deep, structural impacts is not easy, as these effects are often hidden, long-term, and require a systemic shift in how we approach technology. Future work should explore how to address the structural issues in concrete AI cases. Despite the challenges, structural imagination provides a valuable foundation for developing responsible AI.

References

Achterhuis, H. (2001). American philosophy of technology: The empirical turn. Indiana University Press.

AI HLEG. (2019). European Commission's ethics guidelines for trustworthy artificial intelligence (issue May). https://ec.europa.eu/futurium/en/ai-alliance-consultation/guidelines/1

Arnstein, S. (2015). A ladder of citizen participation. In *The city reader* (pp. 323–336). Routledge.

Arnstein S. R. (2019). A Ladder of Citizen Participation. *Journal of the American Planning Association, 85(1), 24–34.*

Benjamin, R. (2019). Assessing risk, automating racism. *Science, 366(6464), 421–422.*

Birhane A. (2020). Algorithmic Colonization of Africa. SCRIPT-ed, 17(2), 389–409.

Blok, V. (2023). *Philosophy of technology in the digital age: The datafication of the world, the homo virtualis, and the capacity of technological innovations to set the world free*. Wageningen University.

Blok, V. (2025). Technology as a theme in phenomenology. In T. Toadvine & N. de Warren (Eds.), *Encyclopedia of phenomenology*. Springer.

Blok, V., & Lemmens, P. (2015). The emerging concept of responsible innovation. three reasons why it is questionable and calls for a radical transformation of the concept of innovation. *Responsible Innovation, 2*, 19–35.

Brey, P., & Dainow, B. (2023). Ethics by design for artificial intelligence. *AI Ethics*. https://doi.org/10.1007/s43681-023-00330-4

Bucciarelli, B. (1994). *Designing engineers*. MIT Press.

Corrêa, N. K., Galvão, C., Santos, J. W., Del Pino, C., Pinto, E. P., Barbosa, C.,... & de Oliveira, N. (2023). Worldwide AI ethics: A review of 200 guidelines and recommendations for AI governance. *Patterns, 4(10).*

Couldry, N., & Mejias, U. A. (2018). Data colonialism: Rethinking big data's relation to the contemporary subject. *Television & New Media, 20*(4), 336–349.

Dignum, V., Baldoni, M., Baroglio, C., Caon, M., Chatila, R., Dennis, L., van Steenbergen, M., Smakman, M., Pavón, J., Micalizio, R., Kließ, M. S., van der Toree, L., Génova, G., Villata, S., Haim, G., Tedeschi, S., Lopez-Sanchez, M., Slavkovik, M., & de Wildt, T. (2018, December). Ethics by design: Necessity or curse?. In *Proceedings of the 2018 AAAI/ACM conference on AI, ethics, and society* (pp. 60–66).

Eubanks, V. (2018). Automating inequality: How high tech tools profile, police, and punish the poor. St. Martin's Press.

Ellul, J. (1964). The technological society (J. Wilkinson, Trans.). Vintage Books. (Original work published 1954)

Franke, U. (2022). First-and second-level bias in automated decision-making. *Philosophy & Technology, 35*(2), 21.

Friedman, B., Kahn, P. H., Borning, A., & Huldtgren, A. (2013). Value sensitive design and information systems. In N. Doorn, D. Schuurbiers, I. van de Poel, & M. E. Gorman (Eds.), *Early engagement and new technologies: Opening up the laboratory* (pp. 55–95). Springer.

Fritts, M., & Cabrera, F. (2021). AI recruitment algorithms and the dehumanization problem. *Ethics and Information Technology, 23*, 791–801.

Gilman, M. E. (2022). Beyond window dressing: Public participation for marginalized communities in the datafied society. *Fordham Law Review, 91*(2), 556.

Goodall, N. J. (2014). Machine ethics and automated vehicles. *Road Vehicle Automation, 93–102.*

Gray, M., & Suri, S. (2019). *Ghost work: How to stop Silicon Valley from building a new global underclass*. Houghton Mifflin Harcourt.

Hagendorff, T. (2020). The ethics of AI ethics. An evaluation of guidelines. *Minds and Machines, 30*(3), 457–461.

Haslanger, S. (2023). Systemic and Structural Injustice: Is There a Difference?. *Philosophy, 98(1), 1–27.*

Hoeksema B. (2023). Digital domination and the promise of radical republicanism. *Philosophy & Technology, 36(1), 17.*

Jobin, A., Ienca, M., & Vayena, E. (2019). The global landscape of AI ethics guidelines. *Nature Machine Intelligence, 1*(9), 389–399.

Korenhof, P., Blok, V., & Kloppenburg, S. (2021). Steering representations—Towards a critical understanding of digital twins. *Philosophy & Technology, 34*(4), 1751–1773.

Kroes, P., & Meijers, A. (Eds.). (2001). The empirical turn in the philosophy of technology. Elsevier Science.

Madaio, M., Blodgett, S. L., Mayfield, E., & Dixon-Román, E. (2022). Beyond fairness structural (in)justice lenses on AI for education. In W. Holmes & K. PorayskaPomsta (Eds.), *The ethics of artificial intelligence in education*. Routledge.

Mills, C. W. (2000). *The sociological imagination*. Oxford University Press.

Muldoon, J., & Wu, B. A. (2023). Artificial Intelligence in the Colonial Matrix of Power. *Philosophy & Technology 36(4).*

Nakada, M. & Tamura, T. (2005). Japanese Conceptions of Privacy: An Intercultural Perspective. *Ethics and Information Technology 7(1): 27–36.*

NLAIC. (2023). ELSA concept. Retrieved 10/04, from https://nlaic.com/en/bouwsteen/human-centric-ai/elsa-concept/

Novelli, C., Taddeo, M., & Floridi, L. (2023). Accountability in artificial intelligence: what it is and how it works. Ai & Society. https://doi.org/10.1007/s00146-023-01635-y

OECD. (2019). Recommendation of the Council on Artificial Intelligence. https://legalinstruments.oecd.org/en/instruments/OECD-LEGAL-0449

Ryan, M., & Blok, V. (2023). Stop re-inventing the wheel: or how ELSA and RRI can align. *Journal of Responsible Innovation*, 1–19.

Ryan, M., De Roo, N., Wang, H., Blok, V., & Atik, C. (2024). AI through the looking glass: An empirical study of structural social and ethical challenges in AI. *AI & Society*. https://doi.org/10.1007/s00146-024-02146-0

Sadek, M., Calvo, R. A., & Mougenot, C. (2023). Designing value-sensitive AI: a critical review and recommendations for socio-technical design processes. *AI and Ethics*, 1–19.

Sax, M., & Wang, H. (co-first author) (2025). From threat to opportunity: Gaming the algorithmic system as a service. Internet Policy Review, 14(2). https://doi.org/10.14763/2025.2.2007

van der Burg, S., Giesbers, E., Bogaardt, M.-J., Ouweltjes, W., & Lokhorst, K. (2022). Ethical aspects of AI robots for agri-food; a relational approach based on four case studies. *AI & Society*. https://doi.org/10.1007/s00146-022-01429-8

Von Schomberg, R. (2013). A vision of responsible research and innovation. In R. Owen, J. Bessant, & M. Heintz (Eds.), Responsible innovation. Hoboken: Wiley: 51–74.

Wachter, S., Mittelstadt, B., & Russell, C. (2021). Why fairness cannot be automated: Bridging the gap between EU non-discrimination law and AI. *Computer Law & Security Review, 41, 105567.*

Wang, H. (2022a). *Algorithmic colonization: Automating love and trust in the age of big data.* UvA-DARE (Digital Academic Repository).

Wang, H. (2022b). Transparency as manipulation? Uncovering the disciplinary power of algorithmic transparency. *Philosophy & Technology, 35*(3).

Wang, H. (2023). Algorithmic colonization of love: Ethical challenges of dating app algorithms. *Techné: Research in Philosophy and Technology, 27*(2). https://doi.org/10.5840/techne202381181

Wang, H., & Blok, V. (2025). Why putting AI ethics into practice is not enough. Towards a multilevel framework. *Journal Big Data & Society*. https://doi.org/10.1177/20539517251340620

Wang, H., Blok, V. & Van Hilten, M. (2025). ELSA Labs for responsible AI: A novel approach for addressing ethical, legal, and social issues. Journal of Responsible Innovation. https://doi.org/10.1080/23299460.2025.2563944

Wang, H. & Blok, V. (2026). ELSA Labs as a method to study AI-based systems at the micro, meso, and macro levels. In S. S. Gouveia (Ed.), The Palgrave handbook on the ethics of artificial intelligence. Palgrave Macmillan.

Winner, L. (1978). *Autonomous technology: Technics-out-of-control as a theme in political thought*. MIT Press.

Winner, L. (1986). *The whale and the reactor: A search for limits in an age of high technology.* University of Chicago Press.

Winner, L. (1992). Citizen virtues in a technological order. *Inquiry, 35*(3–4), 341–361.

Winner, L. (1993). Upon opening the black box and finding it empty: Social constructivism and the philosophy of technology. *Science, Technology, & Human Values, 18*(3), 362–378. https://doi.org/10.1177/016224399301800306. S2CID 145727569.

Winner, L. (1995). The enduring dilemmas of autonomous technique. *Bulletin of Science, Technology & Society, 15*(2–3), 67–72.

Winner, L. (2018). The cult of innovation: its myths and rituals. In E. Subrahmanian, T. Odumosu, & J. Y. Tsao (Eds.), *Engineering a better future: Interplay between engineering, social sciences, and innovation* (pp. 61–73). Springer.

Young, I. (2001). Equality of whom? Social groups and judgments of injustice. *Journal of Political Philosophy, 9*(1), 18.

Young, I. (2011). Responsibility for Justice. Oxford: Oxford University Press.

Zuboff, S. (2019). Surveillance capitalism. A new priesthood. *Esprit,* (5), 63–77.

Chapter 9
Invasive Laboratories: A Critique of Induced Vulnerability in the Retail Atmospherics Literature

Jack Leff

Abstract Laboratories have become an invasive species. In this chapter paper, I critically examine the field of retail atmospherics—an offshoot of marketing and behavioral psychology fields that attempts to engineer emotional atmospheres to induce consumers towards purchasing. Through a close reading of this small, but influential, literature, I analyze the affective politics of retail environments as they struggle to shift from brick-and-mortar stores into digital marketplaces. I argue that the empiricist logics of retail atmospherics have played a direct role in shaping the ways digital marketplaces have experimented and structured themselves, taking advantage of our attention as a way to repackage and sell our subjectivity itself as products. In this way, they extend a laboratory framework into our day-to-day lives, subjecting us to greater surveillance and attempts at manipulation.

Keywords Retail atmospherics · Marketing · Biopower · STS · Behavioral psychology · Affective atmospheres

9.1 Introduction

When you move towards the store, you remember how busy the mall felt as a teenager and how it now seems strangely hollow and barren. However, as you approach the doorway of your preferred clothing store and take your first whiff of that distinctive scent, you're transported back into the store's familiar atmosphere. Inhaling the sense memory of the storefront envelops us in nostalgia. The scent formula has likely changed in the last decade, but somehow the perfumed entryway still smells

J. Leff (✉)
Roanoke, VA, USA

© The Author(s), under exclusive license to Springer Nature Switzerland AG 2026

A. Fox, D. Cressman (eds.), *Arrangements of Power: Tracing Langdon Winner's Legacy Within and Beyond the Philosophy of Technology*, Philosophy of Engineering and Technology 50, https://doi.org/10.1007/978-3-032-17371-3_9

the same to you. Walking between the shelves of the dimly lit store, this olfactory contemplation gives way to memories of aimless wandering. You approach the checkout counter and pay for your purchase as the cashier reads a list of prompts. At the end of the transaction, you're informed that you've been chosen for a survey of customer purchasing habits. You say no, but in the odd temporality of a retail atmosphere this refusal is already too late. Your movements tracked with cameras out of sight, your hesitation at choosing a pair of pants timed, your purchase tabulated as a retail success, your personal demographics already scraped from your credit card, and your very attention quantified. As you wander out of the store, you're more than the purchase you've made. You've been converted into data.

Retail spaces have increasingly become a space of information collection. Customer habits are quantified into data for marketing and managerial professionals to use to make decisions about how to structure retail environments. In this way, retail spaces now resemble laboratories that monitor us, the consumer, with a level of scientific precision. Using surveillance technologies like cameras and attention trackers, storefronts are conducting daily research geared towards efficiency. I'm interested in exploring this possibility and understanding how scientific information is collected in what are nominally thought of as non-scientific environments. What would it mean for our sociological understanding of laboratories if retail and commercial spaces are, themselves, centers for scientific inquiry, and what are the ethical implications of these spaces using unknowing consumers as data sources?

By taking seriously the data collection elements of retail spaces, I raise concerns about the injection of laboratory epistemologies into the fabric of everyday life. I argue that laboratories have become an invasive species, with data collection is part and parcel of necessary daily interactions. Environmental psychologists and managerial studies researchers have joined forces to render retail spaces as sites of intense experimentation in our day-to-day lives. These studies attempt to manipulate the emergent environment in order to produce "atmospheres," textured spaces capable of inducing vulnerability to consumption in the customer (Healy, 2014). Commonly called "retail atmospherics" in the marketing literature, a combination of behavioral psychology and advertising knowledge is brought to bear in order to mediate the consumer's interaction with retail environments. Examining music volume and style, olfactory sensations, visual cues, directional analysis, and other key components of the retail environment that influence the consumer's emotional predisposition to purchasing, researchers try to harvest your reactions to sensory stimuli. As these affective technologies become increasingly fine-grained and powerful, the boundless frontier of digital spaces has been identified as the next logical step in developing more invasive atmospheric tools. To borrow a concept from scholar Nikolas Rose (2007), there has been an attempt by retail atmospherics to make life itself into capitalist substrate. In effect, this would transform the internet into a massive laboratory, ceaselessly skimming data from users. This predatory tendency towards data extraction is so pervasive in our current structure of the internet that not invasively collecting information is often a major selling point for companies like *Mozilla* or a service you can pay a premium for.

This paper examines the relatively small, but influential, interdisciplinary literature on retail atmospherics in order to understand how business and marketing psychology research have been used to develop affective technologies and techniques in retail environments and how this marketing framework influences the politics of technologies. As Langdon Winner writes, "Technologies are never neutral. They are forms of life. In this sense they are also political insofar as they legislate and govern the fundamental patterns which much of modern life assumes." (Winner, 1979, p. 77) As retail atmospherics structure our interactions with consumption then, these atmospheres carry political valances that can be understood through the literature that scaffolds them. By studying this literature then, we can understand the ways that a scientific epistemology embedded in retail spaces reifies dominant norms and political configurations.

First, I will briefly highlight the laboratory dimensions to retail atmospherics alongside Langdon Winner's political work to show how they can shed a light on the technological and emotional entanglements of capitalism. Following that, I review the literature on retail atmospherics and trace the development of empirical methodologies in shaping retail environments. By tracing the transition from brick-and--mortar retail environments to digital ones, I explore the prevailing business narrative underlying the epistemic goals of retail atmospherics research and how it recreates an empiricist, laboratory methodology in an unexpected site.[1] This analysis is necessarily limited, especially in my description of emerging digital technologies. However, I'm hopeful that it will serve as a fruitful entry point into larger discussions about digital retail environments and the intellectual lineage they're developed from. By drawing attention to the strange field of retail atmospherics, I invite other scholars to critically interrogate the ways it has been exported into our day-to-day lives.

9.1.1 Laboratory Studies as Tool for Inquiry

The field of Science and Technology Studies has long concerned itself with the epistemological and political dimensions of laboratory spaces. Foundational figures like Steve Woolgar and Bruno Latour have attempted to understand laboratories as a sociological space that is produced by knowledge and economic factors as much as it produces knowledge that is economically useful (Latour & Woolgar, 1986). One of the central insights of their work, as well as the work of other STS scholars

[1] This distinction between physical and digital is shorthand for the different qualities and considerations of spaces. As scholars Dourish and Mazmanian show, digital spaces are fundamentally material in both their structure and in the circuitry that enables them (2011). More than that though, we're entering a period of commerce where this distinction is increasingly hybridized, and digital spaces are entering more traditional retail environments. For my purposes, I take digital technologies that have permeated retail spaces as fundamentally material in the same way that shelving organization or pumping in of music changes the materiality of the space.

in the laboratory studies space (Mol, 2002), is that laboratories are fundamentally spaces of economic life and as such their epistemological work is tied to the political sphere (Murphy, 2017). Laboratories take on a kind of agency of their own in this light. Through their relationships with these powerful forces, laboratories vie for space and power within the overall mesh of society. This means that the process of data collection itself has the potential to reify existing political norms and raises troubling concerns about laboratories that are ideologically aligned with economic production.

Scholars like Sharon Traweek (1992) have further troubled the confined picture of the laboratory and pointed to the ways science is performed in conferences, email threads, and social spaces in addition to discrete laboratories. In the age of digital technologies, laboratories have only continued to grow. Most notably, with diffuse citizen science research projects that treat the public as a kind of laboratory as they are not only used as scientific labor (Strasser et al., 2018), but citizens scientists themselves are routinely incorporated into data collection and analysis as research subjects (Remmers et al., 2024). In this paper, I am treating retail spaces, both brick-and-mortar and digital retail environments, as similarly diffuse laboratories where hypotheses are tested, variables assessed, data is collected, and empirical knowledge is developed. Unlike in citizen science projects however, there is no consumer buy-in to the project before data collection begins. Instead, the element of data collection on display maps onto what scholar Shoshana Zuboff terms the age of surveillance capitalism (Zuboff, 2020). That is, a shift in economic and retail structures that privileges the extraction of data from the subject in pursuit of the extraction of further profits. I am interested in tracking how retail environments have shifted towards data extraction and what the underlying political goals of that shift entail.

This insight into laboratories mirrors Langdon Winner's insights into the development of technologies more broadly. While he is famous for his analysis of the politics of infrastructure (Winner, 1980), he has long concerned himself with the myriad processes that go into the development of technologies in the first place (Winner, 1992). In his earlier work, he deftly identifies the narrative of "autonomous technologies" that are seen by the public as having a will of their own but actually are moved by socioeconomic forces (Winner, 2001, pp. 11–12, 153). Rather than remove technological agency, technologies on Winner's account both shape and are shaped by the socioeconomic forces that go into their development. This means that technologies have political inertia that can be difficult to overcome but also that interventions can and should be made to ensure technologies serve democratic principles. He argues that interventions prior to the development of technologies are one of the crucial sites where we can mediate the ethics of such technologies (*ibid*, pp. 323–324). It is with this joint insight between Winner's political analysis and laboratory studies' attention to the socioeconomic factors of scientific research that I approach the field of retail atmospherics.

While the bulk of this article will be an immanent critique of the retail atmospherics literature, the larger point I am making is how retail atmospherics as a field offers us insights into the ways retail environments are converted into laboratories and the politics of how that data is generated and used. This is only partially useful, as Winner argues, precisely because it is an established social arrangement that is patentedly unavoidable at this point. We can't avoid retail environments, nor can we evade the surveillance systems that produce data from our encounters with them. However, a more layered understanding of how this data is used can give us insight into ways we can practice resistance to arguably extractive practices and predatory laboratories. In addition to these present-day stakes, I argue that the ways digital platforms like *Meta* are wielding retail atmospherics should cause us alarm as a bellwether for how this kind of affective research will be deployed.

9.2 Retail Atmospherics: A Critical Review

9.2.1 Origins

The term retail atmospherics was coined by marketing scholar Philip Kotler in a 1973 paper in which he examined the different elements that go into producing a retail environment (Kotler, 1973).[2] He writes that retail atmospherics are "the conscious designing of space to create certain effects in buyers. More specifically, *atmospherics* is the effort to design buying environments to produce specific emotional effects in the buyer that enhance his purchase probability" (p. 50). The sounds, aromas, location, architecture, and other details that mediate the customer experience of entering a store are put under the microscope for how they build brand identity and appeal to consumers (p. 51). While Kotler recognized that each element of a storefront does work on its own, what retail atmospherics seeks to describe is the emergent quality of these elements working together. As Kotler writes, "In some cases, the place, more specifically the *atmosphere* of the place, is more influential than the product itself in the purchase decision. In some cases, the atmosphere is the primary product." (p. 48). The emergent sensory phenomenon is the retail atmosphere, and the individual customer takes in the atmosphere as part of the ritual of participating in the retail environment. This means that an antique shop is still an antique shop even if any individual element we associate with an antique shop is changed—i.e., if we remove the musty scent, old-fashioned signage, or grizzled shopkeeper typically associated with antique shops there can still be the emergent sense of being in an antique shop. It's by virtue of the customer inhaling the hallmarks of an antique shop and the emergent atmosphere of the store's organization that the environment of the antique shop is created. His point in the quote above is that we go to stores like antique shops not just because they serve a desirable product, but oftentimes to breathe in the organized chaos of the antique store that makes

[2] In their literature review of retail atmospherics, Lou Turley and Ron Milliman note that interest in the atmospherics of retail spaces can actually be located about a decade earlier (Turley & Milliman, 2000, 193), which shows mostly how engrained these logics are into business geographies.

us think we're a rare or valuable item is around every corner (p. 57).[3] The thrill of possible discovery is part of the experience, and the point of considering retail atmospherics in the design of a business environment is to try and cultivate those kinds of ambient feelings in order to make your store more desirable to customers.[4]

I want to note here that much of the language Kotler uses when developing the concept of retail atmospherics cements racial and gender stereotypes. Certain individuals are not welcome within particular retail atmospheres, meaning that entering a retail environment is always mediated by social norms and politics. As Winner writes in his classic essay "Do Artifacts Have Politics?": "the adoption of a given technical system unavoidably brings with it conditions for human relationships that have a distinctive political cast." (Winner, 1980, p. 128). For Kotler, the set of small adjustments made in a store front that contribute to its retail atmosphere are the technical system and the resulting inclusion or exclusion of populations constitutes the political relations the store has to society. Kotler argues that stores oriented towards women should have an inviting atmosphere that takes advantage of women's "natural" inclination to while away hours shopping, and that one of the constitutive elements of a fishmonger is the "exotic" odor that is implied to dissuade white customers from entry (Kotler, 1973, pp. 55, 58). Advertising has long suffered from racism and sexism, and I want to note that the origins of the retail atmospherics literature are no exception.

Kotler's project here is not purely descriptive, but is prescriptive insofar as he advises managers to play into, and thus reinforce, problematic assumptions. This means that his advice to appeal to women's "natural" shopping proclivities reinforces ideas that women are prone to exorbitant spending or his ideas of "exotic" fishmonger odors are evidence of an olfactory outsider status to nonwhite people. These social norms embedded in the early forms of the work never are examined in the literature, and as a result are worth keeping in mind as we continue to explore how retail atmospherics have evolved. Certainly, there are also ways that spaces cultivate atmospheres of inclusivity/exclusivity that do not reify problematic social structures.

Gay bars, for example, cultivate an atmosphere that is hostile to homophobic patrons, which is a matter of safety for a vulnerable group. That a space can exclude and challenge political norms speaks to the interrelation between space and politics, and perhaps a different direction that retail atmospherics might have been able to travel in at one point in time. In my genealogy of retail atmospherics however, this

[3] Those of you who have discovered a *Broyhill* or *Lane* piece unexpectedly at a steal of a price know exactly what that thrill feels like.

[4] Here I am reminded of Walter Benjamin's concept of *aura* in his classic analysis of the politics of art and the technologies that attempt mimicry (Benjamin et al., 2008). This conceptual tool is used by Benjamin to distinguish between works of art that have captured something of value and can convey an authentic ambiance and the soulless mechanized creations of mass-produced artworks. It has been taken up by scholars of atmospheres in a number of interesting ways (Böhme, 2017; Rauh, 2018), but here is an important reminder of how retail technologies that attempt to engineer an atmosphere are inauthentic and serving capitalist interests.

possibility goes unquestioned by the literature and I think the uncritical nature of Kotler's work is important to highlight as it will be carried through in his assumptions about the relationship between bodies and retail spaces. More than that however, the racial and gendered imagery demonstrates how Kotler is already thinking about the affective interactions between person and retail environment.

According to Kotler, the customer enters the store and is changed by its atmosphere, and by controlling the retail atmospherics the business owner can cultivate desirable consumers and exclude undesirable ones. This is not the kind of bi-directional influence that we see in phenomenological concepts like "intersubjectivity" (Beauvoir et al., 2012; Bergoffen, 1997), but rather a clear desire to have the retail environment influence the affective state of the customer. The goal is to design a space that induces the customer to a positive shopping experience and to exclude customers seen as contaminating the space—the customer is not invited to change the space; the space is generated to engineer the customer and control them. This is accomplished both through the mediation of sensory material like smells, tastes, and sights as well as the exclusionary/inclusionary politics of which bodies are permitted to enter the atmosphere. To return to Winner's work, this is the same way that infrastructure, as a technology that guides human activity, can shape or reshape politics. Winner uses the example of an exclusionary bridge built by Robert Moses to exclude poor people from Long Island beaches due to its restricted height, and while this example is disputed by STS scholars (Joerges, 1999; Woolgar & Cooper, 1999) it nonetheless does an excellent, if apocryphal, job showing how technologies can enforce a political state of affairs (Winner, 1980, p. 124). Similarly, scholars like Andrew Feenberg or Winner, drawing from Lewis Mumford, argue that energy infrastructure technologies can "lock in" particular political formations through the scale of the technology's influence on day-to-day life (Feenberg, 2010; Winner, 1980). Retail atmospherics are a softer touch approach to engineering the consumer than concrete infrastructure; however, the logics motivating the design and implementation remain the same and carry political valences. In the same way that Moses' bridge example is argued to exclude people who can't afford a car to travel to the beach, retail atmospheres have the capacity to exclude some populations while including others. For Kotler then, manager's responsibility is to think about how to cultivate an inviting retail atmosphere that targets *and constructs* the desired audience while also setting itself apart from similar businesses.

9.2.2 The Empiricist Twist in Retail Atmospherics

So far, I have focused on developing Kotler's original formulation of retail atmospherics and how his formulation speaks to the desire to affectively engineer the consumer. However, one striking element of Kotler's work is that his empirical examples are distinctly unscientific. For example, his use of the antique shop as a particular retail atmosphere is meant to evoke an intuitive understanding of the feeling of walking into the antique shop rather than methodically breaking down what constitutive elements ought to go into building that atmosphere. In this way, he offers us a top-down framework that managers are told to build towards rather than isolating variables. So, while his examples are empirical, they are distinctly unscientific in their approach to engineering retail spaces.

The work following Kotler, however, flips the script. Instead of examining the emergent atmospheres, researchers have inverted this way of thinking and tried to understand how to make fine grained tweaks to the retail environment in order to more effectively engender a desire to buy in the consumer. Researchers after Kotler use environmental psychology and managerial studies to develop more effective ways at cultivating retail atmospherics. Where Kotler was interested in retail atmospherics in order to give general, relatively intuitive advice—such as how it's important to cultivate a relaxing atmosphere in high-end furniture stores to offset the anxiety of sticker shock (Kotler, 1973, p. 56)—contemporary literature attempts to use environmental psychology to empirically determine which individual elements of a retail space ought be altered to effectively engineering a vulnerability to consumption (Turley & Chebat, 2002; Turley & Milliman, 2000). Two of the foundational figures for this shift are behavioral psychologists L.W. Turley and Ron Milliman. During the proceedings of the 1994 Academic of Marketing Science conference, they argue that while there has been a proliferation of writing on atmospherics since Kotler's article, the overall lack of coherent methodology creates limitations for the study of the field (Turley & Milliman, 2014). They conclude by saying "Even though researchers in this area are quite certain that the atmosphere does influence retail behavior, there are still a large number of variables and situations that need to be studied. This quest for understanding should mean that atmospheric research should remain viable and important for many years." (p. 83). In other words, they recognize the need for a comprehensive method to address the complex variables that go into producing a retail environment.

The most common framework for retail atmospherics is the Stimulus-Organism-Response model (hereafter referred to as S-O-R), which is a product of 1950s behaviorism (Donovan et al., 1994; Donovan & Rositter, 1982). The central claim of S-O-R is that consciousness and approaching objects are not a pregiven facet of human psychology, but instead mediated by certain environmental stimuli that we can objectively and empirically study through observing individual behaviors (Mehrabian & Russell, 1974). The name is derived from the order of cognitive processes that the model proposes, with the stimuli impacting the organism, who reflects on the stimuli, and then develops a response to the stimuli. This allows for cultivating different reactions to the same object if we change either the social factors surrounding the object or the environmental stimuli that prompted the encounter with the object. Studies adhering to the S-O-R model try to measure this psychological tendency by attempting to isolate individual sensory phenomenon to empirically study the effects of environmental factors on human decision making (Bigne et al., 2020).

By the 2000s, when we see a resurgence in retail atmospherics literature, the S-O-R framework had long since fallen out of favor among psychological researchers (Schnaitter, 1999). Nonetheless, the assumptions buried in its methodology made it perfect for legitimizing and empowering retail atmospherics logic. Not only was it a scientific methodology that attempted to map human behavior through small adjustments to the environment, it also *brought the laboratory directly to retail environments*. Because S-O-R requires studying behaviors in particular environments, storefronts became a welcome and willing partner to expand the boundaries of laboratory research into real world retail spaces. The scientific credibility the framework had and its focus on isolating sensory variables were appealing to researchers and business owners alike. As scholar of predatory gig economies Jamie Woodcock (2020) writes, "Measurement is therefore a key part of managing the labour process" (p. 70) and S-O-R enables the quantification of reactions. For researchers, S-O-R held the promise of increasingly fine-grained ways of introducing psychological manipulation. For business owners, it suggests that attention to sensory details could successfully induce the shopper to consume.

What is lost in this empirical twist is the emergent dimension to retail atmospherics. In Kotler's original formulation, the individual parts of the storefront were important but mostly insofar as they contributed to a cohesive atmosphere (Kotler, 1973, pp. 52–53). This means that while there were empirical case studies, there was not the same appeal to scientific rigor or quantification. In newer research, the individual sensory components take precedence over the emergent atmosphere and typically the atmosphere is seen as preceding the individual components (Grewal et al., 2014; Morrison et al., 2011). Much like a traditional laboratory, the focus is on controlling variables and testing puzzle piece like hypotheses that add up to larger theories about human behavior. However, instead of open inquiry there was a concrete goal that this laboratory was working towards. In other words, the researchers are working backwards from established brand identities to engineer environments that cohesively support the brand by trying to isolate individual sensory components.

Following Winner's propensity for cheekiness, I dub this rhetorical move the "empiricist twist." This is a phenomenon where academic literature modifies their methodologies with a degree of scientific empiricism in order to preserve its legitimacy in the face of a world that increasingly values scientific rigor in all areas of life. More precisely, this is form of boundary-work that incorporates elements of respectable epistemic endeavors to bolster credibility (Gieryn, 1983). However, what makes the empiricist twist interesting to me is the ways that businesses and business oriented academic literatures will employ it as a way to build hype for a product or platform in addition to carving out space for legitimacy. The current attempts to justify AI development with appeals to scientific progress are an excellent example of how empiricist rhetoric is attached haphazardly to a particular scientific or technological endeavor in this way (Bender et al., 2021). Here, this emergent technology is sustained not by a demonstrable ability to generate profit, but instead by investor capital that increases the scale of promise on the back of supposed scientific achievement (Dotan & Seetharaman, 2023). However, at its core, hype-based technologies like AI that trade purely on scientific achievement are sustained by exploited labor (Miceli & Posada, 2022) and surveillance capitalism (Zuboff, 2015).

For retail atmospherics, we can observe an academic interest in how brick-and-mortar stores can restore their brand identity by controlling the micro-climate of the store's world.

Smells, sounds, visual affair, technological gimmicks, emergent brand consistency all become intensely scrutinized by retail atmospherics scholars who adopt Kotler's term in an age of standardized retail experience (Helmefalk & Hultén, 2017; Spence et al., 2014). This research direction has driven the development of increasingly fine-grained ways of manipulating consumers' emotions as they encounter a retail environment. From a review of the literature over the last two decades, it becomes evident of how broadly this intuitive concept is taken up and applied to a multitude of culturally distinct retail environments (see: Rayburn & Voss, 2013; Roggeveen et al., 2020). Moreover, the field of retail atmospherics has begun to reach a much wider audience after taking on this element of scientific rigor. We get a sense of how pervasive and widespread this influence has been by noting just how strongly international the field of retail atmospherics has become, with significant work in the last decade spanning across India, Africa, and Latin America (see: Alakwe & Okpara, 2017; Tanwar & Nangia, 2018; Toldos et al., 2019). There is a smoothing over of cultural difference through an appeal to empiricism on display with an assumed shared international business interest in engineering retail environments.[5]

As critics of behaviorism, such as Donna Haraway, argue, this flattening of cultural difference is intrinsic to the assumptions of behavioral sciences. Frameworks like S-O-R presume a universal human subject that can be understand through ascribing biological motivations to their actions and in so doing culture falls away (Haraway, 2010: 187). Moreover, scholars like Michelle Murphy have pointed out that this universalizing of the human subject reify capitalism and capitalist values under the guise of scientific legitimacy by converting complex sociological behaviors to quantifiable "biological" ones (Murphy, 2017). The goal, then, is to increasingly improve our ability to quantify life itself and in so doing boost productivity and the economic subject. The laboratory mindset behind this work smoothed out cultural differences as variables that can easily be controlled in the milieu of international business and treats the customer as a universal subject whose behavior can, and should, be modified.

That being the case, it should be no surprise that as storefronts become more sophisticated, so too does the study of retail atmospherics. Digital components are added to brick-and-mortar stores (Dennis et al., 2012) that foreground surveillance technologies as a key part of research methodology for customer experience (Hwan & Sergueeva, 2017).[6] Again, we should hear echoes of Zuboff's (2020) concerns

---

[5] Interestingly, this is at odds with Kotler's original examples of the fishmonger stall. Where Kotler reinforces stereotypical ideas of cultural difference, the more recent literature hand waves away the idea of cultural difference in favor of the notion that capitalism defies borders.

[6] Simone Browne tackles the political implications of this increasingly large surveillance network in her book *Dark Matters* and while she focuses on the long history between security, surveillance, and reifying white supremacist notions of blackness, she briefly delves into retail spaces and com-

surrounding the reproduction of surveillance capitalism echo at the prospect of integrating surveillance into ubiquitous storefronts. While the use of surveillance techniques would reach their apogee with the idea of an entirely digital storefront, years before Amazon founded their bookselling platform we see the line between brick-and-mortar and digital spaces begin to blur through technologically mediated retail atmospherics. Screens are now ubiquitous in display fronts, cashiers are timed by their computer systems for how long each customer transaction takes, cameras become embedded into the natural facades of storefronts, and more data are collected about the consumer to yoke demographic information to consumption habits—especially in positions where affective labor is key (Carls, 2007; Dowling, 2007).

All of this data is feeding into the S-O-R framework in an attempt to better understand what stimuli provoke consumer preference. The early connections Kotler draws between brand identity and their customer base are important at this transitionary moment precisely because they speak to the utility of retail atmospherics to mediate the shift into an increasingly digital economy (see: Eroglu et al., 2003; Floh & Madlberger, 2013). The ideas of brand identity and customer subjectivity become enmeshed through the technological interventions introduced into brick-and-mortar stores and this would usher in the logic as consumer-as-product that has come to dominate retail atmospheric design thinking. This shift into hybridized digital and brick-and-mortar storefronts only deepens the work of retail atmospherics through an appeal to empirical methodologies that brings Kotler's ideas into the twenty-first century.

To sum up, the interdisciplinary field of retail atmospherics makes several discrete promises that become achievable by converting retail space into a laboratory. First, they reassure the reader that they can develop ways to induce customers to not only purchase more goods (Akram et al., 2016) but also more specific products tailored to the individual (Muhammad et al., 2014) by manipulating the environment. Second, they make retail atmospheres a question of store identity and promise to make the retail space more competitive in an increasingly competitive commerce environment (Grewal et al., 2014; Roggeveen et al., 2020; Sharma & Stafford 2000). Third, they promise widespread applicability to both the international marketplace as well as the burgeoning digital marketplaces (Koo & Ju, 2010) through technological interventions that lend their legitimacy to behavioral sciences. The goal of this work is to produce consumers, not just attract them, through controlled affective atmospheres (Anderson, 2009) and increasingly fine-grained technological interventions.

mercial spaces such as airports (Browne, 2015). The main point I draw from Browne here is that the inclusion of surveillance technologies can, on its own, reify notions of blackness especially since we know that the way these technologies produce data for retail atmospheric researchers is deeply uneven.

9.3 Digital Retail Atmospherics

The introduction of surveillance technologies into brick-and-mortar is key here as it is the main intervention that retail atmospherics scholars deploy to both "keep up with the times" and further drive the empirical nature of their work. However, the digital revolution promised even deeper integration than surveillance cameras and attention trackers. As digital marketplaces like *Amazon* became a major driver of commerce, the strategies used for brick-and-mortal retail atmospherics would be applied to digital storefronts, including a subsequent massive increase in data collection that would result from platformed social media and shopping cookies.

The resurgence in interest for retail atmospherics is predicated in no small part on the desire by advertisers to develop new atmospheric tools for a shifting informational economy and an emerging "data subject." While much of the hype around data-driven markets is shown by critics to be just that, hype (Hull, 2021), the main parties buying into the hype are precisely those marketers who are interested in retail atmospherics, which means it has strong influence on the development of digital marketing.[7] The promise of retail atmospherics was originally one of spatial engineering through the power of design, aesthetics, and information delivery to customers. These attributes are also the main considerations with user interface of digital marketplaces, which made retail atmospherics a natural place to turn when it comes to adopting brick-and-mortar marketing tools for an information economy. Furthermore, digital commerce has fundamentally different objectives than brick-and-mortar retail spaces. As Winner argues, the defining feature of digital platforms is that they shrink the proximity between seller and buyer dramatically (Winner, 2020). What this means practically is that, in addition to selling you a product, they're also trying to simultaneously capture data about you to sell to advertisers and manage your time. The way they do this is through a reductive move to strip away the vital elements that make data capture harder and streamlining surveillance around an attentional economy. This form of what Schaupp and Diab call "cybernetisation of production" is pushing digital environments to hijack subjectivity at both the consumption and productive levels in order to exercise more fine-grained control over labor forces (Schaupp & Diab, 2020). Importantly, this is an extension of the shift retail atmospherics went from Kotler to behavioral psychology as its main methodology. Now, however, instead of supporting the observational elements the cybernetisation of production has fully reduced the emergent phenomenon of atmospheres to the isolation of individual factors that might influence consumer behavior.

[7] We can also see the unfulfilled hype reflected in companies like *Amazon* halting production on Alexa and other technologies meant to generate profit through data collection. It's doubtful that data collection is going away anytime soon, the concern I raise here is not that we're entering some sort of data economy but rather that data collection presents specific and important ethical challenges regardless of its feasibility at market dominance. We are seeing this come to a head with the hype around the Metaverse technologies that are premised on affective vulnerability to fear of missing out and function as entirely a sales pitch without a product.

The ability to organize a site in such a way that it can induce emotional vulnerability is especially appealing to those platforms whose business models rely on people revealing elements of themselves. Many digital retail environments are, paradoxically, not trying to sell you goods, but rather hook you with an experience akin to Kotler's example of the experience of entering an antique store so that you provide the site's content to other users. We see this especially in major social media platforms and internet infrastructural search engines where you're no longer a consumer, but rather a hybridized consumer-product for advertising firms. The shift between these two digital retail environments is predicated on adopting different elements of the retail atmospherics logic; however, they're brought together by a vested interest in emotional manipulation and as such represent some ethical concerns about the future of retail atmospherics.

9.3.1 Digital Platforms as Data Harvesting Laboratories

The proliferation of data harvesting technologies, which blur the line between physical and digital retail spaces, has necessitated a change in tactics by corporations. Where before retail atmospherics were developed to make consumers more vulnerable to purchasing things, increasingly, these digital retail laboratories are attempting to develop what Michel Foucault calls "confessional technologies" (Foucault, 1990, pp. 59–67) and political philosophers like Gordon Hull argue are structural "disciplinary techniques" in order to operate seamlessly and away from scrutiny (Hull, 2022). These technologies are, in short, the affective tools by which entities compel you to 'confess' details about yourself that comprise your subjectivity (Foucault, 1995). However, in guiding the confession, confessional technologies are also doing the work of shaping your subjectivity. By asking certain questions about who you are (for example, your demographic data or aesthetic interests) and not asking questions about other elements of your identity they can reinforce how you identify along the lines of the confessional questions. In one quick move, you are made to think you're the consumer—which, in a sense, you remain in the eyes of advertisers—while being transformed into the product. At stake in retail atmospherics then is an understanding of the subject and also direct political sovereignty over the right to define the subject. As purchasing habits evolve into penetrating data gazes, the kind of subject being produced is one whose sum total of life is subsumed into capital.

Confessional technologies are required for digital marketplaces because they allow the platform to use your subjectivity as a frontier for capitalist expansion, delving deeper and deeper into your life and, in fact, attempting to create the consumers they want to see in the world. This has been the business model for online platforms that have adopted much of this literature on retail atmospherics, such as *Meta*. Here, we see *Meta* as a platform transformed into a laboratory that demands confessions from its userbase in order to understand how they can construct particular affects. In essence, they're trying to build an atmosphere capable of empirically measuring and controlling affective responses on their platform. It's worth remembering here that atmospheres, despite receiving significant political attention for their ability to connect people, can also be wielded in their networked complexity as a tool of control, especially scientific control (Davies, 2021). The promises of cloud computing and the ability to track ephemeral social networks in real-time by *Meta* forecloses much of that ethical atmospheric possibility (Shaw, 2016). The ultimate goal, as with all of these digital environments, is our attention, and as we continue to consume this content we are drawn to the integrated marketplaces on sites like Instagram or compelled to forfeit our data. This is something Winner observes in the last chapter in the expanded second edition of *The Whale and the Reactor*. Here, Winner notes the prevailing narcissistic and addictive tendencies on platforms like *Meta* that structure the emotional state of the consumer (Winner, 2020). Worryingly, he concludes that these platforms do not simply appeal to these worse angels of our nature, they are in fact engineering these tendencies in ways that threaten democratic norms (ibid). They are reproducing the worst excesses of capitalism through their business model and the kinds of surveillance and affective engineering that model trades on.

The idea that a person's self is room for expansion through affective manipulation is precisely the logical outcome of the empirical retail atmospherics work and presents deep ethical concerns for how we engage online. Instead of using a camera system to surveil consumption habits as in traditional retail atmospherics work, *Meta* has gone a step further by having micro-interactions monitored and categorized to understand consumption habits at a far more granular level. Additionally, the same flaws that I've identified above with retail atmospherics methodology remain and, in fact, are intensified alongside the attempt to turn the consumer into a product through affective engineering (Weinberg, 2020).

We can see the material impact of this limitation in how the platforms of *Meta* structure their social interactions, which is emblematic for social media platforms more broadly.

Rhetoricians of digital platforms like Jennifer Sano-Franchini have done close readings of Meta as a platform. She has traced the impact of their affective work in how they structure their engagement possibilities and the affordances of the platform for allowing us to interact with one another (Fernandes et al., 2022). By focusing on user interface design, Sano-Franchini deftly points to the infrastructural modality of *Meta* and specifically how the infrastructural modality can mediate and encourage a politics of outrage. This is akin to another of Sara Ahmed's observations about the emotional ties that bind online communities of white supremacists (Ahmed, 2015), except the affective logic that Sano-Franchini observes across user interface choices, such as infinite scrolling, are tied directly to the logics of affective power in subject formation (Sano-Franchini, 2018, pp. 389, 396–397).

She points out that it is the extraction of our time, attention, and ultimately ourselves here that motivates the platform and creates the immersive atmosphere that characterizes it as a social media platform (p. 401–402). The extraction of data to produce profit in this way echoes the ways retail atmospheres used logics of inclusion and exclusion to do much of the same, just in a far more sophisticated and controlled manner. In some ways, this is universal across all social media platforms insofar as their design is oriented around profit-making. The atmospheres created in these retail platforms are a logical extension of the choices made by brick-and-mortar stores that seek to maximize profits and social media platforms seek to produce and surveil a kind of subject in ways that are far more granular than we've seen in physical spaces. This necessitates a fundamental shift in the understanding of atmosphere within retail environments, from one that takes the consumers' purchasing habits as the entryway to their subjectivity and towards a more sterilized understanding of affective manipulation via confessional technologies.

This has culminated in the failed project of the Metaverse, which tried to convert your digital avatar into your entire being when it comes to commercial activity. The CEO of *Meta* Mark Zuckerberg's plan for the platform was to have it embody everything from traditional brick-and-mortar retail functions, like grocery shopping, to office work. Through this project, we come full circle back to engineering retail laboratories but this time we are hooked up to a machine that can quantify every micromovement as we explore under the platform of the *Metaverse*. The avatars within this space are intended to mirror every aspect of your life, although the poor execution of these systems have all but scuttled this possibility. As we see a continuing, albeit overhyped, expansion into digital marketplaces we cannot lose sight of the toll of constructing digital retail atmospheres. There is a human cost to the proliferation of these failed interventions into our emotional states. Counterintuitively, the fixation on engineering our emotions ultimately sterilizes our experience of these platforms and attempts to simplify the complexity of being human. To borrow once again from Winner, it is alienating us from not just our work but also the world around us in ways that should fundamentally concern us. By creating a facsimile, it is taking retail atmospherics to their logical conclusion, the reproduction of capitalism and a surveillance economy built on our rapt attention in digital space.

9.4 Conclusion: Resisting Laboratories of Feeling

In the laboratory of capitalism, customers are transformed into empirical fodder, economically and epistemically rendered as singular agents to be experimented upon. Even when we refuse the survey, as in my opening narrative, we are subject to intense scrutiny. When we understand the ways retail atmospherics research transforms the retail space and apply a critical STS lens to study it as such (Weinberg, 2020), the narratives underneath these promises emerge and trouble the hype within these self-ascribed narratives. I understand the so-called transition from physical retail environments to digital retail environments as more than an expansion into a new market, as the literature on retail atmospherics would have us believe, but instead as a troubling rupture in traditional ways of considering marketing, advertising, and commerce. There is a break down in the cohesive narrative this literature is trying to portray and understanding how the emotional technologies and techniques they are developing are being overhauled into confessional technologies is vital to understanding the political stakes of retail atmospherics as an empirical field.

When viewed as a site of experimentation on our atmospheric and respiratory subjectivity, the more insidious elements of these retail atmospheres are made apparent, and we can begin to parse the objectifying nature of their work. Under neoliberalism, public spaces increasingly become monetized through retail atmospherics (Harvey, 2011). Spaces meant for recreation become transformed into outdoor shopping markets and entertainment media are polluted with predatory gambling mechanics meant to prey upon customers. Gordon Hull correctly identify the neoliberal elements intrinsic to certain forms of internet platforms (2021). However, the capitalist promises of these digital retail spaces are fictitious to an extent. Digital marketplaces currently make up roughly 16% of all retail sales, which is substantial growth since the 6% in 2014, but still nowhere near the total dominance desired by retail giants and social media companies (Quarterly Retail E-Commerce Sales, 2023). Yet, the organizational models and supply chain logic attached to digital retail environments have pervaded every other industry. In essence, we see how effective retail atmospherics can be at influencing companies and policy without actually materializing the profit it promises. One could argue that overpromising, changing the world, and then underdelivering is the classic story of neoliberalism, with the hope for endless profit stymied by the realities of a complex social world. The concern remains that the impact of retail atmospherics will continually push corporate research deeper and deeper into affective terrain for the purposes of engineering the consumer.

Winner points to the ways that digital platforms, like *Meta* drives people towards narcissistic traits that undermine democratic ethos of users (Winner, 2020). It is a reshaping of subjectivity through technology that threatens our democratic systems. Even under conditions of failure, when these technologies fail to live up to their hype, the very act of trying to engineer our psyches creates ethical and political harms that ought to be challenged.

These retail spaces as laboratories of feeling can also affect corporate behavior, specifically if consumer trends indicate lost profit for the company. This is evident in the clothing industry, as progressive branding has become increasingly the norm while sweatshop practices remain pervasive. This is not to say that there is a reciprocal relationship. There are massive power asymmetries between corporate influences and the consumer/product. As Winner argues, platforms like *Meta* regulate their systems at the whims of Mark Zuckerberg who is the dominant shareholder and CEO, effectively dictating the platform's functionality and user experience in an oligarchical fashion (ibid). Yet, this exercising of corporate power is not entirely one-sided, as Foucault writes 'where there is power, there is resistance' (Foucault, 1990, p. 95) and there are an endless set of examples of people carving out space for community in the aisles of digital and brick-and-mortar retail environments. What I'm talking about is not the tendency for companies like *Target* to start promoting LGBTQ merchandise, as this is simply another form of retail atmospherics using identity to induce a vulnerability to consumption.[8] While intensely vulnerable to neoliberal cooptation in the form of building their community identity around consumption, there are still noteworthy instances of people using platforms in dissident ways that offer some glimpses in how we might resist neoliberal subjectification.

However, that still leaves open the question of how to address future applications of retail atmospherics that we are more able to intervene directly in. Following in the legacy of critical scholars like Langdon Winner, I argue that the point in interrogating retail atmospherics is to expose their logics and condition the possibility for further disruptions. The predatory market engineering they seek to accomplish has not borne the fruit advertisers have hoped for, but the foray into affective terrain should be concerning for ethicists and political theorists. These are fine-grained, empirical tools that are embedded in the retail spaces we are forced to interact with every day, at least in the United States. They are the pervasive affective atmospheres that seek to not just understand our emotions, but to engineer them entirely. The fascinating thing about the literature on retail atmospherics is the tacit acknowledgement of this fact and the attempts made to build these laboratories of feeling into our daily lives nonetheless. The power of capitalism's logic is on full display here as affective consumption is pushed along above all else, eventually promising to expand into subjects themselves rendering the consumer into the product. This promise has not yet materialized and has been widely critiqued by critical scholars, but my hope is that by exposing retail atmospherics as a literature that helped define these consumptive logics it opens the door to further critiques and a better understanding of capitalism's weaponization of emotional entanglement with our subjectivity.

References

Ahmed, S. (2015). *The cultural politics of emotion* (2nd ed.). Routledge.

Akram, U., Khan, M. K., Hashim, M., Rasheed, S., & Hui, P. (2016). Impact of store atmosphere on impulse buying behaviour: Moderating effect of demographic variables. *International Journal of U- and e-Service, Science and Technology, 9*(7), 43–60. https://doi.org/10.14257/ijunesst.2016.9.7.05

Alakwe, K. O., & Okpara, N. (2017). Influence of retail atmospherics as nonverbal communication on purchase behaviour in the Nigerian retail environment. *Journal of Marketing and Consumer Behaviour in Emerging Markets, 2017*(5), 45–62. https://doi.org/10.7172/2449-6634.jmcbem.2017.1.4

Anderson, B. (2009). Affective atmospheres. *Emotion, Space and Society, 2*(2), 77–81. https://doi.org/10.1016/j.emospa.2009.08.005

Beauvoir, S. D., Simons, M. A., Timmermann, M., & Guilmette, L. (2012). *Political writings* (The Beauvoir series). University of Illinois Press.

[8] And also a useful example of what Jasbir Puar calls "homonationalism," which is the ability for corporate and state forces to coopt progressive veneers in order to reify nationalist politics (Puar, 2017).

Bender, E. M., Gebru, T., McMillan-Major, A., & Shmitchell, S. (2021). On the dangers of stochastic parrots: Can language models be too big? In *Proceedings of the 2021 ACM conference on fairness, accountability, and transparency* (pp. 610–623). Virtual Event Canada: ACM. https://doi.org/10.1145/3442188.3445922

Benjamin, W., Jennings, M. W., Doherty, B., Levin, T. Y., & Jephcott, E. F. N. (2008). *The work of art in the age of its technological reproducibility, and other writings on media*. Belknap Press of Harvard University Press.

Bergoffen, D. B. (1997). *The philosophy of Simone de Beauvoir: Gendered phenomenologies, erotic generosities* (SUNY series, feminist philosophy). State University of New York Press.

Bigne, E., Chatzipanagiotou, K., & Ruiz, C. (2020). Pictorial content, sequence of conflicting online reviews and consumer decision-making: The stimulus-organism-response model revisited. *Journal of Business Research, 115*(July), 403–416. https://doi.org/10.1016/j.jbusres.2019.11.031

Böhme, G. (2017). *Atmospheric architectures: The aesthetics of felt spaces* (A.- C. Engels-Schwarzpaul, Trans.). Bloomsbury Academic, an imprint of Bloomsbury.

Browne, S. (2015). *Dark matters: On the surveillance of blackness*. Duke University Press.

Carls, K. (2007). Affective labour in Milanese large scale retailing: Labour control and employees' coping strategies. *Ephemera, 7*(1), 46–59.

Davies, S. R. (2021). Atmospheres of science: Experiencing scientific mobility. *Social Studies of Science, 51*(2), 214–232. https://doi.org/10.1177/0306312720953520

Dennis, C., Michon, R., Brakus, J. J. Š., Newman, A., & Alamanos, E. (2012). New insights into the impact of digital signage as a retail atmospheric tool. *Journal of Consumer Behaviour, 11*(6), 454–466. https://doi.org/10.1002/cb.1394

Donovan, R., & Rositter, J. R. (1982). Store atmosphere: An environmental psychology approach. *Journal of Retailing, 58*, 34–57.

Donovan, R., Rossiter, J. R., Marcoolyn, G., & Nesdale, A. (1994). Store atmosphere and purchasing behavior. *Journal of Retailing, 70*(3), 283–294. https://doi.org/10.1016/0022-4359(94)90037-X

Dotan, T., & Seetharaman, D. (2023, October 9). Big tech struggles to turn AI hype into profits. *Wall Street Journal*. https://www.wsj.com/tech/ai/ais-costly-buildup-could-make-early-products-a-hard-sell-bdd29b9f

Dourish, P., & Mazmanian, M. (2011). Media as material: Information representations as material foundations for organizational practice. In *Working paper for the third international symposium on process organizational studies*. http://www.douri.sh/publications/2011/materiality-process.pdf

Dowling, E. (2007). Producing the dining experience: Measure, subjectivity and the affective worker. *Ephemera, 7*(1), 117–132.

Eroglu, S. A., Machleit, K. A., & Davis, L. M. (2003). Empirical testing of a model of online store atmospherics and shopper responses. *Psychology & Marketing, 20*(2), 139–150. https://doi.org/10.1002/mar.10064

Feenberg, A. (2010). *Between reason and experience: Essays in technology and modernity*. Inside Technology. MIT Press.

Fernandes, M., Homer, M., & Sano-Franchini, J. (2022, March). Sneaky rhetorics, habitual media, and the affective politics of lurking on Facebook. *Enculturation*.

Floh, A., & Madlberger, M. (2013). The role of atmospheric cues in online impulse-buying behavior. *Electronic Commerce Research and Applications, 12*(6), 425–439. https://doi.org/10.1016/j.elerap.2013.06.001

Foucault, M. (1990). *The history of sexuality: An introduction* (R. Hurley, Trans. Vintage Books ed.) (Vol. 1). Vintage Books.

Foucault, M. (1995). *Discipline and punish: The birth of the prison* (A. Sheridan, Trans. 2nd Vintage Books ed.). Vintage Books.

Gieryn, T. F. (1983). Boundary-work and the demarcation of science from non-science: Strains and interests in professional ideologies of scientists. *American Sociological Review, 48*(6), 781–795. https://doi.org/10.2307/2095325

Grewal, D., Roggeveen, A. L., Puccinelli, N. M., & Spence, C. (2014). Retail atmospherics and in-store nonverbal cues: An introduction: Retail atmospherics and in-store nonverbal cues. *Psychology & Marketing, 31*(7), 469–471. https://doi.org/10.1002/mar.20708

Haraway, D. (2010). *Primate visions: Gender, race, and nature in the world of modern science*. Nachdruck. Routledge.

Harvey, D. (2011). *A brief history of neoliberalism* (1. publ. in paperback, Reprint (twice)). Oxford University Press.

Healy, S. (2014). Atmospheres of consumption: Shopping as involuntary vulnerability. *Emotion, Space and Society, 10*(February), 35–43. https://doi.org/10.1016/j.emospa.2012.10.003

Helmefalk, M., & Hultén, B. (2017). Multi-sensory congruent cues in designing retail store atmosphere: Effects on shoppers' emotions and purchase behavior. *Journal of Retailing and Consumer Services, 38*(September), 1–11. https://doi.org/10.1016/j.jretconser.2017.04.007

Hull, G. (2021, October). The death of the data subject. *Law, Culture and the Humanities*. https://doi.org/10.1177/17438721211049376

Hull, G. (2022). Infrastructure, modulation, portal: Thinking with Foucault about how internet architecture shapes subjects. *Techné: Research in Philosophy and Technology, 26*(1), 84–114. https://doi.org/10.5840/techne2022425155

Hwan, L. S., & Sergueeva, K. (2017). Pictures are worth a thousand words: Photographic narrative inquiry to identify retail atmospherics. *Journal for the Advancement of Marketing Education, 25*, 8–14.

Joerges, B. (1999). Do politics have artefacts? *Social Studies of Science, 29*(3), 411–431. https://doi.org/10.1177/030631299029003004

Koo, D.-M., & Ju, S.-H. (2010). The interactional effects of atmospherics and perceptual curiosity on emotions and online shopping intention. *Computers in Human Behavior, 26*(3), 377–388. https://doi.org/10.1016/j.chb.2009.11.009

Kotler, P. (1973). Atmospherics as a marketing tool. *Journal of Retailing, 49*(4), 48–64.

Latour, B., & Woolgar, S. (1986). *Laboratory life: The construction of scientific facts*. Princeton University Press.

Mehrabian, A., & Russell, J. A. (1974). *An approach to environmental psychology*. MIT Press.

Miceli, M., & Posada, J. (2022). The data-production dispositif. *arXiv*. https://doi.org/10.48550/arXiv.2205.11963

Mol, A. (2002). *The body multiple: Ontology in medical practice*. Science and cultural theory. Duke University Press.

Morrison, M., Gan, S., Dubelaar, C., & Oppewal, H. (2011). In-store music and aroma influences on shopper behavior and satisfaction. *Journal of Business Research, 64*(6), 558–564. https://doi.org/10.1016/j.jbusres.2010.06.006

Muhammad, N. S., Musa, R., & Ali, N. S. (2014). Unleashing the effect of store atmospherics on hedonic experience and store loyalty. *Procedia—Social and Behavioral Sciences, 130*(May), 469–478. https://doi.org/10.1016/j.sbspro.2014.04.055

Murphy, M. (2017). *The economization of life*. Duke University Press.

Puar, Jasbir K. 2017. *Terrorist assemblages: Homonationalism in queer times*. 2 : Next wave. Duke University Press.

Quarterly Retail E-Commerce Sales. (2023). *Census Bureau news CB23-22*. U.S. Department of Commerce. https://www.census.gov/retail/mrts/www/data/pdf/ec_current.pdf

Rauh, A. (2018). *Concerning astonishing atmospheres: Aisthesis aura and atmospheric portfolio*. Atmospheric spaces, no. 4. Mimesis.

Rayburn, S. W., & Voss, K. E. (2013). A model of consumer's retail atmosphere perceptions. *Journal of Retailing and Consumer Services, 20*(4), 400–407. https://doi.org/10.1016/j.jretconser.2013.01.012

Remmers, G., Tzovaras, B. G., Albert, A., Van Laer, J., Wildevuur, S., De Groot, M., Den Broeder, L., Bonhoure, I., Magalhães, J., Assens, S. M., Torrents, E. G., Imre, B., & Covernton, E. (2024). Citizen science for health: An international survey on its characteristics and enabling factors. *Citizen Science: Theory and Practice, 9*(1), 23. https://doi.org/10.5334/cstp.693

Roggeveen, A. L., Grewal, D., & Schweiger, E. B. (2020). The DAST framework for retail atmospherics: The impact of in- and out-of-store retail journey touchpoints on the customer experience. *Journal of Retailing, 96*(1), 128–137. https://doi.org/10.1016/j.jretai.2019.11.002

Rose, N. (2007). *The politics of life itself: Biomedicine, Power, and Subjectivity in the Twenty-First Century. Princeton University Press. ISBN: 9780691121918.*

Sano-Franchini, J. (2018). Designing outrage, programming discord: A critical interface analysis of Facebook as a campaign technology. *Society for Technical Communication, 65*(4), 387–410.

Schaupp, S., & Diab, R. (2020). From the smart factory to the self-organisation of capital: 'Industrie 4.0' as the cybernetisation of production. *Ephemera, 20*(4), 19–41. EISSN: 1473-2866.

Schnaitter, R. (1999). Some criticisms of behaviorism. In B. A. Thyer (Ed.), *The philosophical legacy of behaviorism* (Studies in cognitive systems) (Vol. 22, pp. 209–249). Springer. https://doi.org/10.1007/978-94-015-9247-5_8

Sharma, A., & Stafford, T. F. (2000). The effect of retail atmospherics on customers' perceptions of salespeople and customer persuasion. *Journal of Business Research, 49*(2), 183–191. https://doi.org/10.1016/S0148-2963(99)00004-1

Shaw, D. (2016). Facebook's flawed emotion experiment: Antisocial research on social network users. *Research Ethics, 12*(1), 29–34. https://doi.org/10.1177/1747016115579535

Spence, C., Puccinelli, N. M., Grewal, D., & Roggeveen, A. L. (2014). Store atmospherics: a multisensory perspective. *Psychology & Marketing, 31*(7), 472–488. https://doi.org/10.1002/mar.20709

Strasser, B. J., Baudry, J., Mahr, D., Sanchez, G., & Tancoigne, E. (2018, October). 'Citizen science'? Rethinking science and public participation. *Science & Technology Studies, 32*(2), 52–76. https://doi.org/10.23987/sts.60425

Tanwar, D., & Nangia, R. (2018). Retail store atmospherics in India. *IAHRW International Journal of Social Sciences Review, 6*(10), 2049–2051.

Toldos, M. P., González, E. M., & Motyka, S. (2019). Exploring international atmospherics: The impact on shopping behaviors based on the language of the lyrics of music played in global apparel retailers' stores. *International Journal of Retail & Distribution Management, 47*(4), 368–383. https://doi.org/10.1108/IJRDM-12-2017-0297

Traweek, S. (1992). *Beamtimes and lifetimes: The world of high energy physicists* (1st ed.). Harvard University Press.

Turley, L. W., & Chebat, J.-C. (2002). Linking retail strategy, atmospheric design and shopping behaviour. *Journal of Marketing Management, 18*(1–2), 125–144. https://doi.org/10.1362/0267257022775891

Turley, L. W., & Milliman, R. E. (2000). Atmospheric effects on shopping behavior. *Journal of Business Research, 49*(2), 193–211. https://doi.org/10.1016/S0148-2963(99)00010-7

Turley, L. W., & Milliman, R. E. (2014). Atmospheric research: Past, present, and future directions. In E. Wilson & W. Black (Eds.), *Developments in marketing science: Proceedings of the Academy of Marketing Science*. Springer.

Weinberg, L. (2020). Feminist research ethics and student privacy in the age of AI. *Catalyst: Feminism, Theory, Technoscience, 6*(2). https://doi.org/10.28968/cftt.v6i2.32943

Winner, L. (1979). The political philosophy of alternative technology: Historical roots and present prospects. *Technology in Society, 1*(1), 75–86. https://doi.org/10.1016/0160-791X(79)90010-1

Winner, L. (1980). Do artifacts have politics? *Daedalus, 109*(1), 121–136.

Winner, L. (1992). Citizen virtues in a technological order. *Inquiry, 35*(3–4), 341–361. https://doi.org/10.1080/00201749208602298

Winner, L. (2001). *Autonomous technology: Technics-out-of-control as a theme in political thought*. 9. printing. MIT Press.

Winner, L. (2020). *The whale and the reactor: A search for limits in an age of high technology* (2nd ed.). University of Chicago Press.

Woodcock, J. (2020). The algorithmic panopticon at Deliveroo: Measurement, precarity, and the illusion of control. *Ephemera, 20*(3), 67–95.

Woolgar, S., & Cooper, G. (1999). Do artefacts have ambivalence: Moses' bridges, Winner's bridges and other urban legends in S&TS. *Social Studies of Science, 29*(3), 433–449. https://doi.org/10.1177/030631299029003005

Zuboff, S. (2015). Big other: Surveillance capitalism and the prospects of an information civilization. *Journal of Information Technology, 30*(1), 75–89. https://doi.org/10.1057/jit.2015.5

Zuboff, S. (2020). *The age of surveillance capitalism: The fight for a human future at the new frontier of power*. First trade paperback edition. PublicAffairs.

Chapter 10
On Bullshit Engines' Politics

Damien Patrick Williams

Abstract The contemporary world has become increasingly entangled with so-called artificial intelligence—systems that seem to replicate human speech and thought, proclaiming to make our lives easier, more efficient, and more interesting. But what do these tools and systems actually do, and how do they do it? What systems of power, politics, and other values are embedded within these systems, and who put them there? Throughout this chapter, I take inspiration from and deploy Langdon Winner's theories of how politics are embedded within and reproduced by technologies. Specifically, I use Winner's framing to examine values embedded and entangled with "AI" and how those entanglements lead to what has already occurred, and what can potentially go wrong when we integrate "AI" into various aspects of our lives. This discussion and exploration then form the basis for recommendations as to how we might build these systems differently—that is to say, how we might go about building into these systems the values and politics we want to see rather than the ones we allow to make their way in via our uncritical assumptions and unexamined biases. Neither "AI" qua "AI" nor the kyriarchal, hegemonic, environmentally damaging, capitalist, white supremacist, ableist, eugenic, misogynistic, and otherwise prejudicially biased forms these tools and systems currently tend to take are inevitable. Racist facial recognition, sexist and ableist resumé and hiring tools, and other "AI" systems do not care about the harms they perpetuate or the truth content of their determinations—that's what makes them bullshitters. To change these kyriarchal systems' configurations, we must understand what they truly do, why they are the way they are, what values they reproduce and reify, and how they could be otherwise.

D. P. Williams (✉)
Department of Philosophy, University of North Carolina at Charlotte, Charlotte, NC, USA

School of Data Science, University of North Carolina at Charlotte, Charlotte, NC, USA
e-mail: damien.p.williams@charlotte.edu

© The Author(s), under exclusive license to Springer Nature Switzerland AG 2026

A. Fox, D. Cressman (eds.), *Arrangements of Power: Tracing Langdon Winner's Legacy Within and Beyond the Philosophy of Technology*, Philosophy of Engineering and Technology 50, https://doi.org/10.1007/978-3-032-17371-3_10

Keywords Artificial intelligence · Science and technology studies · Philosophy of technology · Epistemology · Misinformation

10.1 Introduction

One of the main works we'll use to help us think through "AI"'s power, political, and valuative relationships is Langdon Winner's seminal 1980 article "Do Artifacts Have Politics?" Winner was one of the first to articulate a theory of politics and power in technoscientific artifacts which helped us to understand that by the politics of artifacts and systems, we don't simply mean the idea that the political space in which a piece of technology finds itself matters—that the values of the culture in which it's built matter. Those things do matter, but for Winner and many thinking with—as in alongside and after—him, when we say "artifacts have politics," we are also trying to get at an understanding that any built thing also has embedded in it—in the very fact and manner of its construction—some determination about what we think is good and bad in the world.

In "...Artifacts...," Winner primarily showcases artifacts—built things, unique items—but he also discusses systems—interrelated webs of dependencies, networked concepts, and material conditions, including artifacts. Even in the late twentieth century, it was increasingly difficult to distinguish artifacts from systems, but it is especially true today. Within the fields of science and technology studies and the philosophy of technology, understanding the interconnected interplay of artifacts that exist and the webs of systems in and of which they are made is a primary purpose. A dry-erase marker means little without something to write on; the whiteboard means nothing without a marker; the whiteboard's existence entails a particular kind of marker; and the dry-erase marker needs a particular kind of surface. I can write with this marker on other surfaces, like my desk or a piece of paper, but it will not write as well, and it will, in fact, eventually damage the marker. I can write on the whiteboard with other things, such as a sharpie, paint, or a permanent marker—but those things will damage the whiteboard. I can clean the whiteboard with other things than a specialized spray and a cloth—I could use hydraulic fluid, or acid, or I could just splash a whole bunch of water all over the place; and all of those things would work—to an extent. But they would not do the job that ***that*** cleaner was meant to do for ***that*** whiteboard written on with ***that*** marker.

Every artifact has its relation, and those relations are both necessitated by the materiality of the artifacts and the world and contingent on the human choices which have brought them into being. Neither "AI" qua "AI" nor the forms these tools and systems currently tend to take are inevitable; but in order to change the so-often kyriarchal, hegemonic, environmentally damaging, capitalist, white supremacist, ableist, misogynistic, and otherwise prejudicially biased configurations, we must understand what these system truly do, why they are the way they are, and how they could be otherwise.

10.2 Winner's Artifacts

When Winner says that artifacts have politics, he conveys a situation in which a built and physical thing has human values embedded in the very fact of its construction—that in every made thing exists some traces of what its creators think is good and bad in the world. To discuss this, Winner uses the examples of tomato harvesters, signifying labor automation technology in general, nuclear power, and the bridges and parkways of Robert Moses. Throughout the mid-twentieth century, Robert Moses designed much of the iconic landscape of New York City as we know it today. In his comprehensive 1976 volume *The Power Broker*, biographer and historian Robert Caro describes Moses as a racist of multiple stripes and someone who also bore distaste for the lower classes.[1] Moses believed that certain types of people should not be allowed in certain parts of New York City, and he was explicit in communications with others about whom he thought of as worthy of using certain parts of "his" city—Moses did not want Blacks and Puerto Ricans near his most prized public projects. He was also extremely adept at wielding power.

Caro describes how Moses built relationships in which he had bureaucratic authority over things like who got what kinds of contracts to build, roads, bridges, and parks or lay out new building construction throughout New York. He was so good at understanding the law and crafting policy and legislation to suit his needs that he outmaneuvered Fiorello La Guardia, one of the most powerful and politically deft mayors in New York City history. Under the legislation that Moses put together, everything had to run through him. Robert Moses, very specifically, built many of the parkways and the bridges in NYC, and these parkways were intended to be parks that one could literally drive through. A trip on the parkway would be a scenic wooded drive through nature so that rather than just sitting around in Central Park or only driving through smog-filled highways, you would drive through the park, you would get to have a nice Sunday drive with your family in the beautiful scenery of New York City. To this end, Moses designed the entrances and overpasses of the parkways to be of a very specific size, such that trucks or other large commercial vehicles could not drive on them and spoil the scenery. But if trucks can't drive on the parkway, other truck-sized things can't drive there either. And so, city buses could not drive on the parkway.[2]

Many New York residents didn't have personal cars, at this point in time; more specifically, many low-income residents did not have personal cars. Not coincidentally, many of these same low-income residents were of Black and Puerto Rican descent, and they, like most people without a personal conveyance, used the bus to get from place to place. When Robert Moses designed the parkways to be accessible only to cars, only people who could afford to ***own*** cars got to experience the parkway. Correspondingly, when Robert Moses made his bridges and overpasses too low for buses to pass through, and those bridges and overpasses connected the inner

---

[1] Caro, Robert. *The Power Broker: Robert Moses and the Fall of New York.* New York: Knopf, 1974.

[2] Winner, Langdon. "Do Artifacts Have Politics?" *Daedalus*, vol. 109, no. 1, 1980.

city to the beach, the main way to access the beach was by driving there. That, in turn, restricts who the beach is for. If the vast majority of the people using the bus "just so happened" to be the kind of people that Robert Moses deemed undesirable, and it "just so happened" that Moses designed roads and bridges that would manage to exclude just those very people, that, Winner argues, is not just a coincidence: it is Moses' politics coming through in the artifacts and systems he designed and built.

In the many years since *The Power Broker* and "Do Artifacts Have Politics?," people have looked at Winner's work and writing and reexamined the lenses through which historians have interpreted Robert Moses. Some critics have speculated that, "maybe the situation isn't just a cut and dry matter of saying Robert Moses was a racist who built his bridges so that the people he didn't like couldn't get there" even noting that there were other routes busses could take to get to Jones Beach, and other ways for people in the city to get onto the parkway.[3] Others have argued that Moses can be absolved of having made racist bridges because he likely did not do so intentionally.[4] Now, Winner does not argue that every single thing that is built has only one kind of political value to it; rather, his argument is, when we build things, we build our values into it. Consciously or unconsciously, we build our values, perspectives, beliefs, and politics into the things we create. Conversely, our values get shaped by the things we experience and engage with, by the artifacts and systems we interact with daily.

The internet, as a whole, is a massive series of systems—a system of systems. Everything from algorithms to user experience design elements to specific frameworks of architecture and artistic design all exist on the internet and work together to make our experience of the internet what it is. And some of those elements were created so deep into the past, so far away from our present experience, that we simply do not and cannot really know how they shape what we think and what we feel as a result of the systems with which we interact, let alone their constituent elements. Consider the origin of Facebook: a small, knocked-together project called "Facemash," created as a bespoke clone of "Hot or Not"—an online quiz app to rate pictures of women. It was launched in the Harvard intranet system in 2004, the product of a collaboration between Mark Zuckerberg and the Winklevoss twins to make a fun game where people could rate and rank the dates they had. And people ***loved*** it. Facemash takes off amongst the Harvard dorms, and people start to share it and talk about it. It expanded, and people now use it for more than just rating each other. In much the same way, once the internet happened, email happened, and shortly after Facemash reached popularity, people wanted a way to message each other. Somewhere in here, Facemash becomes "The Facebook."

Prior to 2004, a "Facebook" was primarily known as a physical artifact. It was a book in which universities would keep their students' images (faces) and contact information. And so, in being rechristened "The Facebook," Zuckerberg and the

[3] Kessler, Glenn. "Robert Moses and the saga of the racist parkway bridges." *The Washington Post*. 2021.

[4] George Stevens. "Robert Caro's Moses: A Historian's Critique," in Joann P Krieg *Robert Moses: Single-Minded Genius*. New York: Heart of the Lakes Publishing, 1989.

Winklevosses sought to signify their creation of a digital extension of a known category of physical artifact. When people started messaging each other on the Facebook, they started talking about what they might do later on in the day, how they were going to hang out, dates that they might want to go on, or things that had nothing to do with dates; sharing thoughts or early memes—things like chain letters of the kind people would share in an email; all of the usual internet ephemera that happens whenever people get together and do Internet things. While it took time for the Facebook to become Meta as we know it today, it still took its start from the objectification and the rating of women—thus making what we now know about the effects of Meta's products not at all surprising.

Using the theoretical frames partially developed by Winner to understand the social impacts of technology, we can easily understand how a social media technology, which started as a means to objectify and rank women, went on to iterate into a version of itself that drastically harms everyone's mental health, and especially that of women and young girls.[5] We now have multiple studies, external ***and*** internal research, and even sworn congressional testimony, all of which attest to the harm done by the algorithmic content delivery mechanisms in systems like Facebook and Instagram, and now Tik Tok.[6] These internet cultures rely primarily on visual presentation, where comparison to and evaluation against others—objectification and ranking; "hot" or "not"—is a fundamental component. The values and politics of their origins are baked into these algorithmic systems, and they have no easily accessible brakes, making it hard to push back against those values and politics. The affordances of these online ecosystems mean that the easiest way to behave in them is to take the path of least resistance.

Broadly, the affordances of a system or an artifact are those design elements that indicate and influence its use. The affordances of a stool make it easier to sit in than use as a mode of conveyance; the affordances of a website's algorithms and interface make it easier to, e.g., share reposts and comments, than original content or links to other websites. However, a thing's affordances do not foreclose all other possible modes of use; instead, those other modes may simply be harder. But a tool or system may also be what Robert Rosenberger terms "multistable,"[7] meaning such a technology has more than one valence of meaning and experience. Put simply, if we work against the affordances of systems and their attendant harms, we may, to an extent, push the system in a different direction and mitigate those harms, but if

[5] Protecting Kids Online: Testimony from a Facebook Whistleblower Hearing Before the Subcommittee on Consumer Protection, Product Safety, and Data Security of the Committee on Commerce, Science, and Transportation United States Senate, 117th Cong. 1 (2021) (testimony by Frances Haugen).

[6] Protecting Kids Online: Facebook, Instagram, and Mental Health Harm Hearing Before the Subcommittee on Consumer Protection, Product Safety, and Data Security of the committee on Commerce, Science, and Transportation United States Senate; 117th Cong. 1 (2021) (testimony by Antigone Davis).

[7] Rosenberger, Robert. "Multistability and the Agency of Mundane Artifacts: From Speed Bumps to Subway Benches." *Human Studies*. 37, 2014.

we merely let a system do what its design enables, then what it will do are the things that are easiest for it to do, which are what the values built into it will shape it toward. As digital tools and systems change the processes by which we interact with each other and influence the meanings of those interactions, we are thus faced with implications for the social process that shape the production of technology and, more generally, knowledge. Because the production and certification of knowledge is not value-free, it is a value-laden political process, one in which biases and perspectives take part. As Melvin Kranzberg famously put it, "Technology is neither good nor bad, nor is it neutral."[8]

10.3 A (Very) Brief Explanation of "AI"

"Artificial Intelligence" has hype cycles of waxing and waning interest, and all of these fervent conversations about what "AI" should or will be or do, how it will change the world and make things unrecognizable have happened before, and they very likely will happen again. These cycles are referred to as "springs" and "winters." The first "AI spring" happened in the 1950s and 60s, when the concept of "AI" was brought into public awareness for the first time—the idea of a machine that thinks. In the 1960s, a doctrinal rift formed within "AI" research, with one group interested in furthering the development of computation and cybernetics—feedback loops intended to replicate biological processes—and the other working to replicate human neural networks in electronic, or "artificial," form. Both camps, however, disregarded concerns about the impact specific kinds of bodies, emotions, or socialization might have on cognition; these men firmly believed the key to "AI" was to divorce any messy human social factors from the "purity" of abstracted function, rationality, and intellect.[9]

During this same time, "AI" scientists developed language models (LMs)—a method for determining the probable occurrence of words of particular lengths based on context cues such as their starting letter and the word preceding them. One of the earliest examples of public LMs was ELIZA, a program developed by Joseph Weizenbaum at MIT in 1964 and meant to act as a parody of open-ended psychotherapy. Though the program was only a few hundred lines of code, people told ELIZA their deepest secrets as if she were a human therapist. Next came natural language processing (NLP)—a combination of linguistics, computer science, artificial neural networking, and "AI" research concerned with getting computers to interpret, process, and communicate in "human-like" language. In the 2010s, Global Vectors for Word Representation (GloVe) and Word2Vec were among the most well-known NLP models, working by forming statistical correlations to map the

[8] Kranzberg, Melvin. "Technology and History: 'Kranzberg's Laws'," *Technology and Culture* 27. 1986.

[9] Wilson, Elizabeth A. *Affect and Artificial Intelligence*. Seattle: University of Washington Press, 2010.

relationships between words, allowing for an easy way to embed layers of associative meaning. In this way, LMs could represent the semantic connections between words like "woman" and "queen" or "drive" and "car." These early models use so-called machine learning ("ML") to encode their varied English language elements as data and then train the system to meet particular predictive targets and reinforce the associations between the embedded datapoints. Those associations are then mapped as vectors—or mathematical representations—of how strongly they're associated.

However, Word2Vec and GloVe had two very large flaws, the first being that their outputs very often contained obviously prejudicial biases. Again, LMs and NLP are, at base, automated word associations in which massive collections of words and phrases known as "language corpora" are turned into mathematical representations known as "tokens." These systems are then trained on the tokens to predict the ranges of potential associations between them. When you look at the vector space, the word woman doesn't just appear next to the word king or queen, but vice versa—the word "woman" also appears in the vector space very near to the word "secretary," as well as the words "nurse," and "teacher." And "woman" appears very distant from the words "president," "doctor," "CEO." This happens because the collections of natural language on which they were trained was itself packed full to the brim with error prejudicial bias.[10]

Language corpora have to be built, put together out of existing, human-generated, conversational data because the goal in all of this is the reproduction of "natural" language. But datasets for training modeling systems cost money. To this end, language sets such as the Enron Corpus represent treasure troves of large amounts of free or low-cost public domain natural language interactions; but they are also very often filled with human beings speaking in bigoted, immoral, or even just unconsciously biased. And this is because the Enron corpus is a collection of more than 600,000 emails generated by 158 employees of the Enron Corporation in the years before the company collapsed, all of which entered the public domain as the Enron Corporation was on federal trial for public corruption and defrauding their investors. The Enron Corpus contains people trading and rating pictures of women; slurs against people of Muslim background; stereotypical "jokes" about the sexual proclivities of Black and Asian peoples; and more. Due to this, we find that language modeling systems which use this training data replicate, iterate on, and exacerbate the built-in biases and perspectives of the data from which they are trained. And so we find what initially seems like surprisingly gendered bias, racial bias, or ableist bias within the system—and yet, it should not, in fact, be surprising. Because those systems are trained on data—on human interactions—which themselves contain those biases. This also results in outcomes such as automated résumé sorters

[10] Caliskan, Aylin; Bryson, Joanna J.; Narayanan, Arvind. "Semantics Derived Automatically From Language Corpora Contain Human-Like Biases." *Science*. 2017.

rejecting the applications of women and certain minorities at higher rates than white men and, more recently, rejecting any résumé that mentions disability in any way.[11]

The second problem with Word2Vec and GloVe was that they couldn't map associations across larger collections of text; in fact, the number of associations they could make actually decreased as the amount of text increased. The solution to this problem wasn't simply adding more and more associations or tokens to parse; this problem space required a different way to parse the tokens themselves. And so we come to the transformer framework.

The "GPT" in ChatGPT stands for "Generative Pretrained Transformer," a phrase describing a system of interoperable algorithms used to weigh, arrange, sort, and return large associations of language. The transformer process is the overarching operational algorithmic framework enabling everything from GPT text-based transformers for OpenAI, to Google's Gemini, to Midjourney, to deepfakes, and large language models (LLMs). Their training is about creating a greater fidelity of statistical correlation between text and text, text and image, text and video, or text and audio. To do any of this work, you first use the model in an untrained way, and that untrained result is then compared to the desired output. It is then retrained toward the desired output, and then that is put into practice to generate new results. It's a feedback system. Human beings then engage in a process known as reinforcement training. That reinforcement process oftentimes requires a large group of humans to teach these systems whether they've gotten it right or whether they've gotten it wrong and then nudge them closer and closer towards getting it right. That's what reinforcement means in this space, but the language corpora that are used for training still must come from somewhere. For LLMs, the language corpora used the internet as a whole.

As the name suggests, LLMs are larger versions of LMs; in fact, the data that LLMs use can be in the billions, or now even trillions, of large tokens. Ideally, these tokens must be labeled in the "right" way, but what does "right" even mean in this context, and who are the people doing the training which enables all of this? Because this is how biases get embedded within GPT and LLM systems: The data on which they are trained comes from sources with those biases encoded. Over the past several years, this has become a standard understanding throughout LLM systems: biased training data gets biased results. But many people don't know that the biased training data that we're dealing with is, in many cases, decades old, as the old, prejudicially biased corpora get scraped and re-scraped every time the outputs from a system trained on them is scraped. And so, these newer LLM and GPT-type "AI" systems retain the same problems of prejudicial bias as we find in GloVe and Word2Vec—but now those problems have iterated faster and become more interwoven with every aspect of our lives.

---

[11] Kate Glazko, Yusuf Mohammed, Ben Kosa, Venkatesh Potluri, and Jennifer Mankoff. 2024. "Identifying and Improving Disability Bias in GPT-Based Resume Screening." In Proceedings of the 2024 ACM Conference on Fairness, Accountability, and Transparency (FAccT '24). Association for Computing Machinery, New York, NY, USA,

While the above is not a full history of "AI" or of all of "AI's" "springs" and "winters," it is meant to emphasize that the social values present in each turn have emphasized the automation of speed, efficiency, and the mimicking of human rationality. These values have remained present, even when other, previously disregarded mental and social process elements have been engaged and brought into the "AI" space. It could be argued that we are in the middle of a protracted and segmented "AI" spring, in which both the definition of "AI" and what gets defined ***as*** "AI" change more rapidly and within an ever-shorter timeframe. However, it could be said that the underlying framework characterizing all of these technologies is the automated algorithmic system, a term simultaneously ubiquitous and (often intentionally) opaque. An algorithm is merely a set of instructions—standardized operations within which particular weights and measures can be adjusted to change the outcome of a process. But when you alter an algorithm, you must alter every element of the whole—the framing context—to ensure the final product still turns out correctly. And if this is not done intentionally and carefully, then what comes out on the other side may be very messy indeed.

Ultimately, the algorithms that enact LLMs and GPT-type systems do not care about giving you answers which correspond to the facts of the world. What these systems do is statistically model the most-likely-to-be-acceptable result. LLM and GPT-type "AI" are, quite simply, trying to tell you a story that will jibe with your preconceptions and biases, based upon the inputs you've given the system; this is not the same as "providing facts" or "telling you the truth." Conversely, when the system generates something that does not correspond with reality—a process anthropomorphically described as "hallucinating"—it isn't "lying" to you or "making a mistake;" in fact it's doing the same thing as it did in providing you an answer that ***did*** corroborate facts about the world. And in epistemology and philosophy of language, we have a very specific word for a speech act that is not at all concerned with falsity or truth.

When we talk about the truth content of speech acts, we are talking about something very specific. If I seek to tell you the truth as it is, I am seeking to tell you what I believe corroborates with the facts of the world as I know (or believe I know) them. If I am mistaken about my beliefs, then I haven't lied to you; I either had an error in my own understanding or in my relation of my understanding to you. To lie to you, I have to seek to deceive you, and to seek to deceive you, I have to tell you something which ***I believe*** runs counter to the facts. To do that, I must know—or believe I know—what those facts are. But what if I don't care about telling you facts or truth, and I don't care if you're deceived by what I have to say? In that circumstance, what I'm saying isn't the truth, and it isn't a lie—what I'm saying is bullshit.

10.4 The Politics Which Frame "AI"

In Harry Frankfurt's 2005 monograph *On Bullshit* he describes the process of "bullshitting" as telling a story and not caring whether it conforms to the facts of the matter (is true), or whether it does not (is false).[12] And for Frankfurt, this is the most potentially harmful kind of speech act because there is no goal for the bullshitter other than telling a good story. And if you're an irascible musician spinning yarns about trains and circus performers, then telling a good story is part of the job. But if you're trying to have a conversation with other people about shared public values, or trying to design a system in which people can make decisions about their daily lives, then a foundation of bullshit is a very dangerous thing indeed. LLM and GPT systems are always making things up, but sometimes what they make up happens to correspond more closely to the facts of the world, and sometimes it doesn't; either way, the "AI" doesn't care. "AI" systems do not care about facts; more to the point, they ***cannot***, so far as they are currently constructed, care about facts. These "AI" systems only care—insofar as they are able "care" about anything—about generating a response that is statistically the most likely to be accepted by the prompter based on the correlational values of their input tokens. This, in turn, means that there is no way to remove "hallucinations" from these systems—because the "hallucinations" fundamentally are what transformer models does. They are bullshit engines.[13]

And Google's Sundar Pichai, Apple's Tim Cook, and OpenAI's Sam Altman have all admitted this fact; these CEOs have fully admitted that there is no way to eliminate "hallucinations" in practice and that their systems cannot help but bullshit you. You can mitigate the bullshitting, and you can minimize it, but you will never stop an LLM or GPT-type system from, at some point in its operation, fabricating wholecloth something that does not necessarily conform to reality because it figures you might like that answer best. And that "AI" system will present that fabrication to you with the same confidence and certainty as it would any fact that ***does*** conform to reality. This has major implications for the places in which generative "AI" is already interwoven with our lives and understanding this requires first reckoning with the concepts via which we discuss them. As we've discussed, even the term "AI" contains assumptions about the nature of intelligence (or consciousness), about what parts of "human-like" cognitive processes can be replicated in machine processes—but the term also contains and enacts assumptions who or what can have intelligence at all.[14]

---

[12] Frankfurt, Harry. *On Bullshit.* New Jersey: Princeton University Press. 2005.

[13] Williams, Damien P. "On Bullshit Engines: The Socioethical and Epistemic Status of GPTs and other AI." Lecture in the Forum in Ethics, Law, and Society Lecture Series, hosted by the Sonoma State University Philosophy Department's Center for Ethics, Law and Society. 2023

[14] These assumptions have no agreed-upon criteria, and a not-uncontroversial history of application; for more on this see Williams, Damien Patrick. *Belief, Values, Bias, and Agency: Development of and Entanglement with "Artificial Intelligence"*. Diss. Virginia Polytechnic Institute and State University, 2022.

In 2022, the Georgetown Center for Privacy and Technology announced they would no longer refer to systems as "artificial intelligence," "AI," or "ML," arguing that these terms obscure and often outright mislead the lay public more than they illuminate. I agree, and also argue that this terminological obfuscation facilitates the hype-laden corporate marketing of those trying to develop and sell "AI" systems and also enables circumstances wherein even supposed critics nonetheless ***uncritically*** reiterate the overblown claims of the technologies' creators. This comingled hype thus allows "AI" proponents to claim that they've built something (a) extremely powerful and (b) over which they have no responsibility or control. But the basic fact of the matter is that we don't know what "intelligence" is, let alone what we mean by "AI," and so the term has merely become a marketing tool rather than pointing to anything real or concrete. "AI" is a moving cipher seeking a referent.

Simultaneously, this willful occultation of what happens inside these systems further confounds and oppresses the marginalized human beings most often subjected to them. This, too, is why all instances and variants of "AI" are here present in scare quotes or otherwise qualified: To better critically and productively trouble both this occultation and the assumed notions of both "artificiality" and "intelligence." Elsewhere, I also use this to highlight and directly challenge the unwillingness of so many to confront the religious and magical valences of their own thinking about "AI." For instance, when "AI" systems determine what targets to hit in military conflicts, the technology is often given religiously valanced names. In Gaza and Lebanon, the Israeli military uses a system known as "the Gospel."[15] When we use religious language to discuss technology, it taps into a conceptual paradigm that signifies certainty, even for those who might consider themselves atheists. It provides a framework of guarantee and expertise that says, "This is trustworthy, this is correct, this is infallible. It is the gospel."[16]

This happens in much the same way that technologies that rely on code and math garner veneers of "objectivity," which in turn gives people a way to wash their hands of any responsibility for what those very human technologies do: the repeated habit of and inclination toward this process is what Meredith Broussard terms "technochauvinism."[17] This happens not only with GPTs but with "AI," algorithms, and code more broadly as well. "It is math, it is code. It can't have biases in it. Can it?" And yet have long known that it can—that technological systems can have human perspectives embedded in them.[18] And while these systems are increasingly

[15] Mariam Barghouti, "The Gospel: Israel Turns to a New AI System in the Gaza War." Al Jazeera. 2023.

[16] For more detailed discussions of the religious values and perspectives embedded in "AI," see Williams, Damien Patrick. "Any sufficiently transparent magic…" *American Religion* 5.1 (2023); and Williams, Damien Patrick. "Religious Belief and Practice in the Age of 'AI,'" hosted by the Emory University Candler School of Theology. 2024.

[17] Meredith Broussard, *More Than a Glitch: Confronting Race, Gender, and Ability Bias in Tech.* Cambridge, MA: MIT Press. 2023.

[18] Porter, Theodore M. *Trust in Numbers: The Pursuit of Objectivity in Science and Public Life.* Princeton, N.J.: Princeton University Press. 1995.

applied in theater of war where a decision is a matter of life and death, as we've discussed above, it can also be about life and death at home.

Neurodivergent individuals, non-native English speakers, women, and racially minoritized individuals who tend to "mask" or seek to "pass," in their communications, be they spoken or written. From the outside, this process often appears like someone over-thinking or being overly formal or precious—even "stilted" or "robotic"—with their word choices. In written form, those phrasing patterns get flagged as "anomalous" by GPT checkers, thus designated as less likely to be "human." This is done not only by automated systems but also by other humans. As we'll discuss below, many people begin to believe that "we can always tell" when utterances are produced by "AI" systems—even when they're not. When we rely on "AI"—when we certify "AI" systems as arbiters of truth, as determiners of fact—what we are doing is giving these systems the power to make judgments for us and safeguard knowledge. In so doing, we are imputing heuristic capabilities to generative "AI" systems that they do not possess. And when we begin as a culture and as a society to rely upon them, we begin to reflexively seek to enact the kinds of judgments that we believe an "infallible" system might make. Thus, we begin to believe that we can see the "markers of AI" where they do not exist.

Telling people who they are rather than heeding their lived experiences is a way of enacting power and control over how people perceive themselves and their world. In this same way, institutions and organizations further concentrate their influence and power via circumventing or manufacturing the consent of their users and those made subject to their systems. When "AI" companies suggest the need for nearly fantastical advances in nuclear power or the retasking of existing sources as a way to redirect attention away from their astronomical water use and carbon footprints, it is in an attempt to control the conversation about the human cost of "AI"'s creation, deployment, and operations.[19] When "AI" companies claim to make major advancements in the fields of machine vision or reinforcement learning, but have actually just offloaded that work, some of which has some of the most dangerous effects on mental health of any job on today, to exploited workers in the majority world, those are also attempts by those corporations to control the conversation.[20]

And this is why "AI" CEOs such as Sundar Pichai or Sam Altman, and corporations ranging from Facebook to Zoom, and now even academic publishers like Wiley, Oxford, and Taylor and Francis try to claim the right to scrape anything they want from the internet in general and our creations in particular without our consent

[19] Verma, Pranshu, and Shelly Tan. "A Bottle of Water Per Email: The Hidden Environmental Costs of Using AI Chatbots." *The Washington Post*. 2024; "Microsoft deal propels Three Mile Island restart, with key permits still needed," Reuters. 2024.

[20] Perrigo, Billy. "Exclusive: OpenAI Used Kenyan Workers on Less Than $2 Per Hour to Make ChatGPT Less Toxic." Time.com. 2023; Zeff, Maxwell. "Amazon Ditches 'Just Walk Out' Checkouts at Its Grocery Stores." Gizmodo. 2024; Birhane, Abeba, Sanghyun Han, Vishnu Boddeti, and Sasha Luccioni. "Into the LAION's Den: Investigating Hate in Multimodal Datasets." Advances in Neural Information Processing Systems 36 (2024).

and use it for "AI" training data.[21] When these people and groups tell you "we can't do our job if we can't use copyrighted material," they are both lying to you and trying to get you to accept a particular way of thinking and operating within the framework of these technologies. When Meta partners with Ray-Ban to make facial recognition-enabled smart glasses that they put on the market and then place that product in every single commercial break of every television or online streaming show you watch, they are trying to lock you into a way of thinking about these technologies. However, these corporations' and organizations' preferred frameworks are, in fact, not necessarily the only options, though they want you to believe that they are. It is not inevitable that these tools, that these systems, are built in the ways that they are built, but believing in their inevitability allows the people who make the systems to tell us how it then must be from that point forward, if they're inevitable, then we have to build them. And if "we" build them, then our competitors can't build them—or at least "we" have the most influence over how they're built and what tools and capabilities are available within these systems. "If American companies build AI systems," the argument runs, "then the US controls the market, not China;" or "If we make these chips in Ohio, we don't have to worry about what hidden firmware may be hustled into the system."

All of this is a process of paradigmatic capture. When you capture the paradigm—the way of thinking about, researching, building, deploying, and using something—from top to bottom, you then determine how everyone else engages with it from that point forward. However, there are a whole host of people and groups working to regulate and alter how we build "AI" systems, including executive orders and laws from the United States and new "AI" acts from places like the EU and China. China, for instance, just decreed that there must be a way to digitally watermark everything that comes out of an "AI" system or that said system cannot be used in China.[22] Digital watermarking and other means of provenancially verifying "AI" outputs was proposed as early as 2022, floated as a means to know what is and is not "AI," and at that time (and even more recently) "AI" companies said it would be completely infeasible.[23] Since then, it has been revealed that not only ***can*** it be done, but since nearly their inception OpenAI has had a system to steganographically embed a marker to identifies text generated by their "AI" with 99.9% accuracy, absent large-scale editing on the system's product by a human.[24] This can

[21] Kwon, Diana. "Publishers are selling papers to train AIs-and making millions of dollars." *Nature* 636.8043 (2024)

[22] Yang, Zeyi. "China's Plan to Make AI Watermarks Happen." 2024. *Wired*; China Law Translate. "Measures for Labeling of AI-Generated Synthetic Content (draft for solicitation of comments)." 2024.

[23] Francis, Matthew R. "The Ethics of Artificial Intelligence-Generated Art" SIAM News. 2022; Hague, Douglas C., et al. Comment #NTIA-2023-0005-1215, National Telecommunications and Information Administration Request for Comment on AI Accountability. 2023. https://www.regulations.gov/comment/NTIA-2023-0005-1215.

[24] Seetharaman, D., and M. Barnum. "There's a Tool to Catch Students Cheating with ChatGPT. OpenAI Hasn't Released It." *The Wall Street Journal* (2024).

be done via simple word correlation choices, which can be weighted within the model, and the fact that they are not widely enacted until companies are forced to do so indicates something about those companies motives.

10.5 The Politics Enacted Through "AI"

We are now beginning to see generative "AI" in all the places where algorithmic systems have long been interwoven into our daily lives. We are continually confronted with headlines highlighting some new horrific outcome of "AI," yet people and institutions seem to still be rushing full steam ahead to use it. For instance, during this election cycle, we have seen the use of generative "AI" deployed from and intended to influence political perspectives, from "AI" generated Taylor Swift and her fans being used to claim endorsements for multiple candidates to "AI" generate audio of candidates' voices meant to convince people not to vote. Left unaddressed, the danger that generative "AI" poses to the civic processes is truly massive—but not wholly unprecedented.

In addition to the already-mentioned sexist and ableist prejudicial bias displayed by résumé sorters, researchers have also found that systems entangled with generative "AI" bullshit engines recommend white male job candidates almost 100% of the time.[25] "AI" image generators or "enhancers" consistently create racist and sexist images.[26] Generative "AI" systems which are being used to guide people through getting disability benefits replicate the same kinds of ableist biases that we've seen in previous older versions of the system. Racist and sexist biases show up in regards "AI" systems' recommendations for who should get which kinds of health care, and even how an "AI" summarizes someone's healthcare interactions.[27] Because facial recognition systems don't see dark skin tones well, and the "AI" systems that are used to sort through surveillance videos are trained on and operating out of repositories of data rife with misidentified Black people, Black people comprise the vast majority of misidentifications, arrests, and erroneous prosecutions by police of people based on "AI"-backed systems of facial recognition.[28]

---

[25] Stifler, Lisa. "AI Overwhelmingly Prefers White and Male Job Candidates in New Test of Resume-Screening Bias." GeekWire. 2024.; Wilson, Kyra, and Aylin Caliskan. "Gender, race, and intersectional bias in resume screening via language model retrieval." *Proceedings of the AAAI/ACM Conference on AI, Ethics, and Society*. Vol. 7. 2024.

[26] Ananya. "AI Image Generators Often Give Racist and Sexist Results: Can They Be Fixed?" *Nature*. 2024 Mar; 627(8005); Elizabeth Laraki, *Tweet*. http://web.archive.org/web/20240000000000*/https://x.com/elizlaraki/status/1846252781851890026

[27] Koenecke, Allison, Anna Seo Gyeong Choi, Katelyn X. Mei, Hilke Schellmann, and Mona Sloane. "Careless Whisper: Speech-to-Text Hallucination Harms." In *The 2024 ACM Conference on Fairness, Accountability, and Transparency*. 2024.

[28] Sanford, Alyxaundria. "Artificial Intelligence Is Putting Innocent People at Risk of Being Incarcerated." The Innocence Project. 2024.

Combined with generative "AI" policing integrations such as police body cameras with generative text capabilities for writing reports or making predictive frames for resource distribution.[29] The datasets for training such a system will consist of the tagged and parsed collection of often handwritten, post-hoc recollections police reports, reports which are themselves written based on the responding officers' notoriously and prejudicially biased memories.[30] So, while some researchers suggest the use of "AI" tools as a means to review bodycams and check ***against*** officers' memories—possibly even to help determine what patterns and markers might be present in interactions where in which police are more likely to harass people—it is baffling to hear anyone defend any "AI" as "objective" or "impartial."[31] It is especially concerning when that assertion receives no pushback from the interviewing journalist.[32]

Generative "AI" integrations given over 6000 sample loan applications based on data from the 2022 Home Mortgage Disclosure Act recommended denials for Black applicants more often than statistically identical white counterparts, recommended higher interest rates for Black applicants, and labeled Black and Hispanic borrowers as "riskier," than white counterparts, and results showed that white applicants with low credit were approved 95% of the time, with Black applicants approved less than 80%.[33] These outcomes can be traced to the prior orders of both algorithmic digital and original analog redlining, so the recurrence of it in new technologies should not come as a surprise.

At least equally alarmingly, there's recently been a raft of "AI" generated mushroom foraging guides available for sale online[34]—sometimes wholly generated images based on what the system has been trained a mushroom is statistically supposed to look like, while other guides retrieve existing images and then generate text to describe them by means described above. Shortly after they integrated image search capabilities, I had Google Bard (now Gemini) list out ten wild mushroom species to forage, and recipes to go with them; it correctly identified some, but the ones it misidentified had errors ranging from strange preparation suggestions to outright describing toxic species as edible—complete with tasting notes. Now, when someone asks for a bedtime story about a geese in the woods who had a delicious dinner, a statistically-more-likely-to-be-accepted response is fine; but when a

[29] Cox, Joseph. "Here Is What Axon's Bodycam Report Writing AI Looks Like." *404 Media*.2024.

[30] McIlwain, Chalton. *Black Software: The Internet and Racial Justice, from the AfroNet to Black Lives Matter. New York: Oxford University Press.* 2019

[31] Kaste, Martin. "Human Reviewers Can't Keep Up with Police Bodycam Videos. AI Now Gets the Job." NPR. 2024.

[32] Ibid.

[33] Gross, Paige. "As AI takes the helm of decision making, signs of perpetuating historic biases emerge." 2024. Utah News Dispatch; Bowen III, Donald E. et al., "Measuring and Mitigating Racial Disparities in Large Language Model Mortgage Underwriting." 2024. Available at SSRN 4812158.

[34] Trybuch, Elan. "AI generated mushroom books – life and death." New York Mycological Society. 2023.

real person is feeding their family based on a bullshit engine's advice, that is a different story Because a fun thing about mushrooms is that many of them appear very similar, and sometimes what you think is a tasty dinner ingredient is actually a poison that will liquefy your liver; you will die, and it will hurt the entire time that you are dying.[35]

In the realm of academia, we have seen increasing evidence that "AI detectors"—systems that themselves use generative "AI" to purportedly identify text created ***by*** generative "AI"—contain prejudicial bias against both non-native English speakers and neurodivergent students, a thus urging us to further consider differences in how expectations and perceptions of gender and race can affect people's communication.[36] Some people's lived experiences are least likely to be adequately accounted for in either the training data or the weighting architectures of GPTs or other "AI" tools, the implications of which are well exemplified by incidents related by Rua M. Williams: One in which a non-native English-speaking student used ChatGPT to rewrite their work to "sound more white," and another in which Williams was falsely accused of using "AI" for their own writing was. [37] Both of these instances exemplify the normative pressures within both academia and "AI" research.[38]

And yet, even though we recognize all of their varied harms, these systems are still in use, which makes it seem like it is more valuable to have a system that does not work well and which harms marginalized and minoritized individuals and groups than it is to do the work necessary to fix these systems. Again: One can alter how the "AI" generates text or images, thus producing outputs identifiable by reference to the generating system—this is inherent to how LLM and transformer systems function. It may take extra time and require a bit more effort from, thus costing their parent companies more—and these are reasons why these kinds of processes are said to be "infeasible" or "impossible," right up until the moment we learn they have been done. Thus, why we have to be on guard for that moment when "AI" companies shift from saying safeguards are "impossible" to complaining about their difficulty and expense. "AI" companies, like most other corporations, absolutely could protect their users' rights, dignity, autonomy, and the value of their creativity and labor—they just don't think the users are worth it.

---

[35] Erden, A, et al. "Acute Liver Failure Caused by Mushroom Poisoning: A Case Report and Review Of The Literature." Int Med Case Rep J. 2013 Nov 22;6.

[36] Williams, Damien Patrick. "Bias Optimizers." *American Scientist* 111.4 (2023); Liang, Weixin, et al, "GPT Detectors Are Biased Against Non-Native English Writers," *Patterns*, Volume 4, Issue 7, 2023; Davalos, Jackie and Leon Yin. "AI Detectors Falsely Accuse Students of Cheating—With Big Consequences." 2024. *Bloomberg*.

[37] Williams, Damien Patrick. "Disabling 'AI': Biases and Values Embedded in 'Artificial Intelligence.'" *Handbook On the Ethics of Artificial Intelligence*. Edward Elgar Publishing, 2024; Williams, Rua M. Fractal Echo. https://kolektiva.social/@FractalEcho/111949020988947948, https://kolektiva.social/@FractalEcho/110740599615481285.

[38] Maiberg, Emanuel. "Scientific Journals Are Publishing Papers with AI-Generated Text." *404 Media*. 2024; Williams, Damien P. "Scholars are Failing the GPT Review Process." *Historical Studies in the Natural Sciences* 54.5 (2024)

Clearly, whether we're talking about GPTs and LLMs or just "machine learning" and algorithms, "AI" systems definitely have politics. The problem is not and never has been that what "AI" is and does is "out of step" or "unaligned" with human values, but rather that the values with which "AI" is aligned are (a) very, very human, (b) woefully under-examined, and often (c) terrible. Those values embedded here tend to be racism, sexism, ableism, efficiency, power-over, speed, control, white supremacy, eugenics, hegemony, kyriarchy; what they don't tend be are prosocial consensus reality, intersubjective truth, or facts. Because obfuscation, misinformation, and disinformation are valuative positions—and they are also mechanisms of control.

10.6 Epistemic Crisis as Power

When we think about how we construct systems of knowledge, we are often concerned with ideas of objectivity and the determination of fact; but we can also consider intersubjective knowledge and consensus reality—making a shared understanding of our world together.[39] Both Lorraine Daston and Peter Gallison's 2007 book and Jonathan Grunert's 2019 dissertation discuss the history of how we come to think about what is and is not objective truth.[40] When we think about an anatomical drawing or nature drawing of a leaf or a taxidermy representation of an animal, what is the goal of that representation? Is it a perfect re-inscription of one particular animal, one particular plant? Or is it a way of trying to capture something about all forms of that animal, that plant? This is a debate that's been alive within image-making for centuries and must now be extended to what we are seeking to represent when we use generative "AI." Are we seeking to represent the factual truth of the matter, or are we seeking to represent, in a broad way, the generally accepted perceptions of things as they have tended to be recognized by people?

Because what we do every time that we use a generative "AI" system is point to the past: Every single "AI" system that exists uses data ***from*** the past. It is trained on things that have been, and yet it is used to teach us something or to try to teach us something, about what might be—a task for which they are fundamentally not suited. If you want to look at the way that people thought about race, gender, and sexuality up until even 2021, then "AI" is the system for you. But if you want something that can tell you how things might be tomorrow, or how things ought to be 10 years from now, that is not what these tools are for. Expectation and PR hype shape

[39] Williams, Damien Patrick. "Constructing Situated and Social Knowledge: Ethical, Sociological, and Phenomenological Factors in Technological Design," in *Engineering and Philosophy: Reimagining Technology and Social Progress*, edited by Guru Madhavan, Zachary Pirtle, and David Tomblin; Springer, Cham, 2021.

[40] Daston, Lorraine, and Peter Galison. *Objectivity*. New York: Zone Books, 2007; Grunert, Jonathan D. "Strict Fidelity to Nature: Scientific Taxidermy, US Natural History Museums, and Craft Consensus, 1880s to 1930s." (2019).

what the public thinks "AI" tools are, can be, what they should do, and how people use them. Reflexively, people's expectations and use then shape how the researchers, designers, and companies tell the story of how we should understand "AI". Here, we can think about the lessons of the social shaping of technology or social constructivism, which say that though we embed our values in technology, we cannot simply think that reshaping technology will completely reframe our values.

The problems of technochauvinism mean that we are more likely to trust as verifiable that evidence which technology provides. When Google changed the online search process, people began to treat positions in the ranked results as markers of trustworthiness; however, researchers like Safiya Noble and others have shown that trust and veracity had little to do with what we see in our searches. [41] To be clear, the certification of knowledge or the possibility of anything like an "objective truth" is by no means uncontested. Theorists in STS, epistemology, and philosophy of technology have demonstrated that the situatedness of time, place, and culture, from laboratories to peer review, can all have major impacts on what we know and how we know it.[42] But many people still seek certainty and treat technologically-enacted search results as a means to it, and they are thus vulnerable to over-trusting mathematically and technologically enabled tools—integrating generative "AI" into search makes all of this worse. We cannot Technofix our way out of values problems; and when we try, weird stuff happens. When Google started pushing their "AI Overviews" into search, there were immediate problems with the results, ranging from the absurd to the downright dangerous.[43] The pushback online was swift, and yet google changed very little about the process other than the things it wouldn't discuss. At the time of writing, the "AI" overviews are still in place, enacting terrible values through their operation.[44] When Google reframed their Bard LLM system as "Gemini" and started integrating audio, video, and image generation, somebody who was trying to prove a point about "wokeness"[45] prompted the system for an image of "racially diverse Nazis."[46] And so, Google's Gemini system generated a picture of racially diverse Nazis. Because that is what it was asked to do, and that is all these systems do: Try to provide you for what you asked for. If people place their trust in "AI" generated images as being just the same as facts, we risk shaking or even outright corrupting the project of shared knowledge and consensus reality and

---

[41] Noble, Safiya U. *Algorithms of Oppression: How Search Engines Reinforce Racism.* New York: New York University Press, 2018.

[42] Cf. Madeleine Akrich, Bruno Latour, Steve Woolgar, Karin Knorr Cetina; Williams, D. P. "Constructing Situated and Social Knowledge."

[43] Cole, Samantha. "Google and Amazon AI Say Hitler's Mein Kampf Is 'a True Work of Art'." 404Media. Mar 6, 2025.

[44] Gilbert, David. "Google, Microsoft, and Perplexity Are Promoting Scientific Racism in Search Results." *Wired.* 2024.

[45] A term originating in Black American communities to indicate savviness about structural racism, but which has since been co-opted to indicate a kind of performative and exaggerated social justice mindset.

[46] Gilliard, Chris. "The Deeper Problem with Google's Racially Diverse Nazis." The Atlantic. 2024.

corrupting the processes for certifying knowledge. AI generated images are not, on their own, a way to create a more equitable or more just society—believing that they can result in deeply offensive results that do more harm than good.

If an image is presented with a story as though the image ***is*** that story, exactly as it happened, and that image turns out to have no basis in reality, then we should question the story itself; that is, if the supplemental evidence used to certify a narrative is found to be false, then we should not automatically treat the story as true. Otherwise, the veracity of every conjunction of statement and evidence is assessed only by our feelings about them—by the vibes. And if we ***want*** everything just to be determined by vibes—by confirmation of our preexisting biases within bespoke realities—then that's okay, but we ought not try to call this a process of figuring out "truth" or "facts" through "evidence." We should just admit we like how an image, story, or narrative assertion makes us feel, no matter how spurious. Without at least that level of honesty, all we're doing is bullshitting each other at a speed and scale heretofore unheard of, and frankly, that will just keep degrading our ability to trust and communicate with other people. When generated images get called out as "AI" and members of the public respond with, "ok yes the image is probably A.I. generated ***it doesn't mean the story isn't real***" (emphasis added), then it can credibly be claimed that we are in a moment of cultural epistemic crisis.

Multiple papers in the past 2 years demonstrate that over-reliance on supposedly objective "AI" tools diminishes critical thinking capacity and prevents students from building the kinds of foundational skills which allow them to learn more complex concepts, adapt to novel situations, and grow into experts.[47] That lack of expertise and capacity has a direct impact on people's ability to discern facts, produce knowledge, and even participate in civic/public life. The diminishment of critical thinking skills makes people more susceptible to the abovementioned dangers of propaganda and other forms of dis- and misinformation. This further erodes trust in shared knowledge, consensus reality, and even expertise, thus exacerbating and worsening the next turn on the cycle when it starts all over again. And all of this creates the very conditions by which authoritarians seek to enrich themselves and cement their control: by obfuscating their intentions, offloading responsibility onto the systems, and undercutting the mechanisms which can empower the populace to understand and oppose dictators, theocrats, fascists, and kleptocrats.

Our reactions to new technologies are often slow, because the ways we think about them are uncertain. They are new, unfamiliar, and so we don't necessarily have a framework in which we can fit them; but many times, this slowness is intentional. How we think about these things are directed by whoever gets to have a conversation about them, and the people who drive these conversations, in turn, get to shape the cultural narrative about the thing. And so, given that we cannot but have

[47] Gonsalves, C. "Generative AI's Impact on Critical Thinking: Revisiting Bloom's Taxonomy." *Journal of Marketing Education*. (2024).; Bastani, Hamsa, et al. "Generative AI can harm learning." *Available at SSRN* 4895486 (2024). Lee, Hao-Ping Hank, et al. "The Impact of Generative AI on Critical Thinking: Self-Reported Reductions in Cognitive Effort and Confidence Effects From a Survey of Knowledge Workers." Microsoft. (2025)

values embedded in our technology, we have to take a close look at the values we want embedded in our technology to determine whose values and which values we want embedded in our technology. We will not be able to have an unbiased "AI." We will not be able to have a completely value-free, value neutral technological system. It is not possible. We are human beings, and everything we touch, we put our perspectives into. Our artifacts and our systems have politics, and our other values too. So, what values do we ***want*** to have in the things that we create?

None of this must be this way. There are so many points when the prevailing paradigm and general milieu of the "AI" project could have been changed; we didn't have to be here. And we don't have to stay here. However, this will not be solved by using the same frameworks of consent violation, theft, and manipulation in a "public" or "open source model," but rather by reframing to a model that respects the dignity, autonomy, rights, and freely given and meaningfully informed affirmative consent of those whose information gets used as training data, and on whose populations the systems are deployed. The US Federal Communications Commission moved very quickly to make illegal the use of "AI" generated audio in campaign ads. The "Joe Biden" robocall went out January 9, 2024; the FCC ruled in February; and by late September 2024, the person who created the ad was fined $6 million .[48] Taylor Swift specifically noted that part of her reason for publicly endorsing Vice President Kamala Harris was that Swift felt her own autonomy, rights, and consent were violated by "AI" claiming to speak for her.[49] And after public outcry, Zoom had to change their plans, as did Microsoft for their "Co-Pilot" rollout.

Change can be affected when the public is educated and mobilized on these issues at multiple levels. And so we need multiple voices working together to try expanding not just the technical education about, but our understandings of the social and human implications of "AI" itself and everything that falls under that very nebulous, empty cipher of a header.

These systems are being widely implemented, and those willing to sow doubt and use confusion to control others will profit more and faster than those who hold to consensus, prosocial norms.[50] In the run-up to university finals season for Spring 2025 OpenAI began offering two free months of ChatGPT, marketed specifically to students; Donald Trump signed an executive order mandating the use of "AI" in K-12 education; and that same administration began feeding reams of sensitive personal data into "AI" systems, even as they cut funding in technoscientific research.[51]

---

[48] Shepardson, David. "Consultant fined $6 million for using AI to fake Biden's voice in robocalls." Reuters. 2024.

[49] Torek, Belle. "Taylor Swift, Kamala Harris, AI Falsehoods and the Limits of Counterspeech." 2024. Tech Policy Press.

[50] Chesney, Bobby, and Danielle Citron. 2019. "Deep Fakes: A Looming Challenge for Privacy, Democracy, and National Security." California Law Review 107 (6); Schiff, Kaylyn J, Daniel S Schiff, and Natalia Bueno. 2023. "The Liar's Dividend: Can Politicians Claim Misinformation to Evade Accountability?" OSF.

[51] Lee, Cydney. "In First Big Marketing Push, OpenAI Makes ChatGPT Free for College Students." AdWeek. 2025; Trump, Donald J. "Executive Order Advancing Artificial Intelligence Education

Norms like diversity, equity, and inclusion, consensus reality, and the processes of shared knowledge-making are under attack, precisely when we could benefit from them. We need to be but are being expressly prevented from considering the situated knowledge and intersectionality of who gets to shape everything from the political sphere to the technical training of the people who build our technologies. Many public policy, research, and grassroots organizing groups are working on ideas for new technologies and new ways to build these technologies, so how do we bridge the different needs and perspectives of the people doing this work?

Suppose we want technological systems that do not depend on predatory, extractive, kyriarchal logics which don't replicate biases against gender, disability, race, or other aspects of identity—which don't first hustle those harms in under the guise of the next big thing and then disseminate them to different places. To undo and prevent these harms, we must work very hard to educate our students, our communities, and ourselves about how these systems work, what they can and cannot do, what impacts they have, and how we might work together to build something more overarchingly just in their place.

for American Youth," April 23, 2025. Larkin, CJ. "100 Days of DOGE: Assessing Its Use of Data and AI to Reshape Government." TechPolicy Press. 2025; Garisto, Dan "Exclusive: NSF stops awarding new grants and funding existing ones." Nature. 2025.

Chapter 11
We Are All Now the Whale and the Reactor, We Are All Now the Coalfields

Crystal Cook Marshall

Abstract Crystal Cook Marshall reopens Langdon Winner's *The Whale and the Reactor*'s title chapter as a theoretical, philosophical, and political warning that we have not heeded. We are no longer site-specific whales swimming near nuclear reactors. Coalfields, often a focus of Cook Marshall's work, now bring risk to us all, dispersing and excelling catastrophe—we create coalfields, and we are coalfields. Blending personal narrative, excavation of STS, and direct encounter with Hurricane Helene, Cook traces our contemplation of "natural" terrors (the whale) to an industrial terror (a nuclear reactor) to our manufacture of nature into an industrial terror which offers no one and nothing any harbor.

Keywords Science and technology studies (STS) · Climate flickering · Rural industrialism · Anthropogenic climate change · Coalfields · Environmental catastrophe

Chapter 10 of Langdon Winner's *The Whale and the Reactor,* Second Edition shares the title with the book, yet it remains under-analyzed and under-utilized in comparison to the previous chapters.[1] In this penultimate passage, Winner juxtaposes the natural world and the risks of a nuclearized world. He also calls on us to search for the limits, to assert the limits. As I envision him in his contemplation of the whale and the nuclear reactor along the Pacific Coast, I want to find a way to heed Winner's

---

[1] Winner, Langdon. *The Whale and the Reactor: A Search for Limits in an Age of High Technology.* 2nd ed. Updated edition. Chicago: University of Chicago Press, 2020.

C. C. Marshall (✉)
College of Agriculture and Environmental Sciences, North Carolina Agricultural and Technical State University, Greensboro, NC, USA
e-mail: cacookmarshall@ncat.edu

© The Author(s), under exclusive license to Springer Nature Switzerland AG 2026

A. Fox, D. Cressman (eds.), *Arrangements of Power: Tracing Langdon Winner's Legacy Within and Beyond the Philosophy of Technology*, Philosophy of Engineering and Technology 50, https://doi.org/10.1007/978-3-032-17371-3_11

call. I wonder if the call still stands. Am I right that he is asking us—when have we had enough? When is enough, enough?

Why juxtapose a nuclear reactor and a whale? Why then and there? In a tone personal and intimate, Winner describes his visceral experience and reaction to the siting of a nuclear reactor and his sighting of a whale on that same coast. The siting and sighting contiguous, he frames these two massive objects in tandem, the reactor, twentieth century's preoccupying risk—nuclear power (meltdowns) and nuclear war (world annihilation). The whale represents a nineteenth century relationship to risk—localized, immediate, deadly—yet something we could pursue, master, dominate. In the late twentieth century, the nuclear reactor and its meltdowns and radiation producing genetic or fallout monsters filled our collective imagination. The whale represented a previous era's concept of the monster of nightmares and novels.

The 1970s, the decade preceding the publishing of *The Whale and the Reactor*, had been a decade of rehabilitation for the whale's reputation from Leviathan to a creature to be celebrated and protected.[2] By the time *The Whale and the Reactor* was published, biologist Robert Payne had put out his recording of whale song, and the whale as energy, as a resource, as fertilizer, was transformed into a creature to be marveled and saved.[3] During the peak of the nuclear push post World War II, the sentiment could be captured as, *so what if a nuclear reactor put whales at risk? Technology is progress. Why do we even now need whales?* Payne brought us their voices, their conversation. He opened the query—*what might the whales say to us? And shouldn't we be listening?*

By the 1970s and 1980s, in a burgeoning and broader environmental movement, nuclear annihilation and poisoning represented all that was wrong with what we humans could achieve and inflict. In elementary school, we children practiced hiding under our desks in case of Soviet nuclear attack.

Popular movies in the theater and on television filled my generation's imagination with how earth's annihilation (or being saved from annihilation) would go, or what life would be like after nuclear war: *Wargames*, *Red Dawn*, *Mad Max*, *The Terminator*, *The Day After*.[4] Depictions often fell in line with Biblical themes of the Apocalypse. Life will be one way, we humans will be going along, maybe filled with hate or hubris, but then *total annihilation will come*. All at once. Swiftly. This will flare and extinguish in a massive and bright flash. The aftermath will be immediate and clear.

During the crisis of nuclear war, presidents would act valiantly. The populace would mobilize. Enemies would be clear and targeted. Avoiding nuclear war, its occurrence, or responding to its aftermath would command all our attention, national reserve, resolve, and resources. There would or could not be an American enemy from within. Response would be unanimous and total. The news channels and cycle

---

[2] Rothenberg, David. *Whale music: Thousand mile songs in a sea of sound*. MIT Press, 2023.

[3] Suter, Keith D. 1981. "The International Politics of Saving the Whale." *Australian Outlook* 35(3): 283–294. https://doi.org/10.1080/10357718108444757.

[4] Bartlett, Andrew. "Nuclear warfare in the movies." *Anthropoetics* 10, no. 1 (2004).

would respond in unison and perspective. Rural, suburban, and urban Americans would stand together and fight side by side. Or, at the very least, we'd have good-looking anarchy, ride on wild-looking technomobiles with medieval spikes, wear high fashion leather biker gear, possess superhuman strength, zoom into a Thunderdome with a scantily clad Tina Turner and an aptly tanned and rugged, shirtless Mel Gibson.

As I originally conceived of this chapter, I envisioned pulling the metaphor of Winner's whale and reactor through to relate it to the relationships between industry that externalizes costs and environmental clean-up in rural places and hides the spoiling from city folk and the spoils from rural folk. Considering the *Whale and the Reactor*, we usually parse Winner's discussion of Robert Moses and politicized highways/bridges in New York City, the deliberate construction of our totalitarian workplaces, and the intentional creation of technology to undercut union organizing. In these respects, Winner stands foundational to me in my examination of the totalitarian work arrangement, especially of rural places where work for one large employer or for one industry is the only game in town. Elsewhere, I have argued that in these kinds of places, the totalitarian workplace arrangement extends into one's living space and life outside work. In brief, in a city, if you can't work in one kind of industry, you can arrange work in another. In a single sector rural space, often there is no other industry to transfer into, and, if you are known as a rabble-rouser, other businesses may not hire you as that would be an affront to local political power in which the industry is embedded or to which local politicians or the populace is beholden.

I have spent a lot of time examining how these relationships play out in the space of a town or municipality and then extrapolating from that the rural industrial space as a type, a kind of "Big Rural," where single sector industries dominate swaths or corridors of rural places, and thus dominate their people and politics.

Complexity, intensity, and idiom are my fortes. It can take me a while to convert those into our currently dominant preferred academic or theoretical, measured communication style of precision, brevity, hesitance or reserve, and lack of metaphor. I am going to lay out for you more specifically the circumstances in which I now find myself sorting out what it means to do Science and Technology Studies (STS), and, then I am going to hammer the square peg of Winner's whale and reactor metaphor until it wedges somehow into the round hole of these circumstances. I am also going to try to take risks from the nineteenth century and twentieth century with me.

For the last 14 years I have been living and working in Central and Southern Appalachia in the United States engaged as a social change practitioner and researcher, and most recently as applied research faculty in Cooperative Extension at the country's largest public Historic Black College or University (HBCU). The mission of the university where I work explicitly addresses the needs of underserved farmers both urban and rural, and of rural people of color more broadly. Our most recent institutional plans of work also explicitly name climate change as a focus of our collective research and services.

In the last 6 months, not equivalent, but both significant, events have occurred that impact how I conceive of conducting STS scholarship and the work at my university. The first is my on-the-ground experience of Hurricane Helene in the US State of North Carolina. The second is the massive dismantling of major portions of the Fifth Branch, State-sponsored expertise and agencies, in the United States.

Prior to the latter, for this chapter, I had written: *the work of STS scholars now is to trace the totalitarianism in our pockets as well as climate flickerings and destruction. It would be hard to convince me that there's anything more important to think or write or act about right now for anywhere, urban or rural.*

As I revise this chapter, the US government is eradicating federal support for climate change research and adaptation and emergency response, as well as for broad swaths of medical research. Only this morning a friend, an oncologist, that does cancer treatment lost her research funding, a year from having results on definitive new treatment for leukemia. In the case that a current grant I have also gets cancelled, I am currently working with the tech transfer department at my university to find venture capital to fund the analytic platform my research team and I have in development for farmers to make decisions about climate change and their personal and farm practices. I have been sent several lists of words and concepts we can no longer use when we write research grants or talk about university-affiliated research publicly. The case study county I have worked in for years in West Virginia has recently flooded and for the first month thereafter, the government did not send any funds or assistance to address clean up, health, food, logistics, roadways, etc. A neighboring coal county, once abandoned for development, is now crawling with data center proposals. The top 20 officials in West Virginia experienced in emergency preparation and response have either taken early retirement or been fired, and the current person running this system there has never managed a public emergency of any sort. Last week, the US government dismantled the US National Climate Assessment, dismissing some 200 scientists.

In STS, we know how to classify and catalog and examine the processes of examination and cataloging: We are losing our jobs to automation. We are losing our minds to algorithms. We are losing our world, our Whale, to anthropogenic climate change that could have been prevented if a major change of course had become the central focus of our government as with a war effort, or sending a man to the moon, or to Mars. We are already losing/have lost the war on adaptation. We are losing/have lost our collective US practice of democracy in a co-production with the destruction of science and technology achieved through a campaign of dismantling, conspiracy, and deflating social trust.

There were warnings that we would enter echo chambers, but we did not know we would take artificial intelligence into them with us, to manufacture as much nonsense as we possibly could tolerate, to test just how far we were willing to entertain the manufactured, how far we would go in the Milgram experiment.

The United States also no longer serves as a certain check on nuclear threat. And so on the list goes.

To state the obvious, we were in a different ground situation and different world when Winner contemplated the whale and reactor.

What is the work of STS in the United States now beyond cataloging the destruction of science? Are we left only excavation of policy and power and proposals?

How do we meet Winner's call to action now?

While November 5, 2024 started a new era for science in the US, for those of us in Central and Southern Appalachia, the world and our scientific understanding of it already had changed irrevocably September 27–29 with Hurricane Helene. Before then I thought I had known what I wanted to say about our human relationship to technology—that Winner's nuclear reactor's location out of sight aligned with how I understood massive technologies in the rural: they keep the lights on but we don't see them; we are purposely uninformed or unaware of the risk they pose. I worked through aligning rural places with the whale.

Hurricane Helene continues to defy what I can imagine. It also defied and continues to defy scientific classification.

It opens the question that what if we have never imagined the destruction that is to come, the destruction that is here now?

Although Hurricane Helene is out of the news cycle, it does not make it any less momentous. In its aftermath, the algorithm on social media sent me this description by firsthand witness Terri L. Dolan. It offers as good an introduction to Hurricane Helene as any I could muster:

> Throughout all of the 200k years Humans have been on Earth, a VERY small portion of us have witnessed Geological events and lived to tell the story. This is a club I hope none of my family, loved ones, offspring or enemies ever have to witness. The ppl of WNC [WNC = Western North Carolina—a mountainous region of the Southern Appalachians and the Hurricane's epicenter of destruction] are now a member of a 5% club and that means we all have hard truths to rectify.
>
> I feel compelled to attempt to try and explain the difficulties the human brain has in trying to process being in this 5% club. Please be patient with me while I try.
>
> Most Geological events happen over a longer period of time, like for instance global warming or the melting of Glaciers. Our event started and stopped (as far as we know) over the span of 48 hours.
>
> Most ppl in the 48hr period only witnessed one of the above underlined causes from a Geological Event. A few folks witnessed two at the same time. I haven't yet met anyone who was unfortunate enough to witness all three in the same 48Hr period but If I do I imagine it will be a first responder.
>
> Our eyes are designed to see things in 2D, with the help of the outdoors and our other senses we can then use that information to build a 3D model in our minds. Anyone who has witnessed our Valley in person over the last 6 weeks can attest to the fact that photos do not do justice to the devastation.
>
> One of the most difficult things for me is driving in the Swannanoa Valley. As the earth's dirt (landslides) have been cleared off our roadways I now use more back roads to get to the destinations I need to reach. There is not a day that has passed since the storm that I do not break down into tears while driving.[5]

Here I pause to find that hammer to smash a square peg into the round hole.

See, the technology I have studied most closely is that of coal. I have studied it as a system of stakeholders. I have studied what its promise of production had

---

[5] Bradleigh Cameron Miller. Facebook post in *I Grew up in North Carolina* (group), November 12, 2024. Accessed November 16, 2024. https://www.facebook.com/groups/1494018708056447.

meant. In the nineteenth century, coal fueled "progress," enabling factories, trains, ships, heat, electricity—not only propelling the collective of humanity into new relationships with technology, but also with consumption, environmental capacity, and even the nature and experience of light and dark.

Hurricane Helene is the closest I have come personally in terms of what *swift* nuclear annihilation was supposed to bring. Of course, while maybe of a type, each tragedy is unique. 200,000 people did not die in this like in the nuclear bombing of Japan. Yet, the scatter and scale of climate change claims more than that many lives a year around the globe.[6] In historic time, this attack may seem as if it is also happening swiftly, a different conceptualization of all-at-once—an attack diffused, dispersed, waged through emissions, through dirty bombs filled with coal. Climate change *so far* is more similar to Pope Leo VIV's recent assertion that with so many conflicts happening at once around the globe—we currently have a world war happening in pieces.[7] Anthropogenic climate change grabs each live creature by the throat and slowly squeezes.

Maybe Winner ought to have sat next to a different kind of energy plant. Maybe he ought to have contemplated the title *The Mountains and the Coal-Fired PowerPlant*. In STS we are also people of our times; we also chase the latest risk, the latest technology. Granted, it is hard to catch up with the latest and most monstrous much less remember to remember still relevant Frankensteins.

How much of STS is now consumed with artificial intelligence and machine learning? And yesterday it was consumed with gene editing and nanotech. Doing STS on coal or agriculture or even on autos or food waste and dozens of other still relevant Frankensteins is just not sexy.

For sure, Mr. Latour, we have never been modern. We are still not yet caught up to the risks that exploded from the nineteenth century.

With more time I could catalog so many more ways that yesterday is still today—from techno optimism to political trends to religion to yellow journalism to coal barons and tech barons and power.

I could catalog how the energy-fueled apocalypse is not like in the movies.

I know coal. I am a product of its promise, and the risks we missed, its risk to us all. Let me set more of this stage.

According to the International Monetary Fund, global subsidies for fossil fuels reached a record $7 trillion in 2022.[8]

---

[6] World Health Organization. "Climate Change and Health." *WHO*, last modified October 30, 2021. Accessed April 30, 2025. https://www.who.int/news-room/fact-sheets/detail/climate-change-and-health.

[7] Pope Leo XIV. "Pope Leo XIV Says He Fears WW3 Had Started 'in Pieces' as He Urges 'Just & Lasting Peace' in Ukraine." *The Sun*, May 11, 2025. https://www.thesun.co.uk/news/34904812/pope-leo-ww3-fears-ukraine/.

[8] Simon Black, Antung Liu, Ian Parry, and Nate Vernon, *IMF Fossil Fuel Subsidies Data: 2023 Update*, IMF Working Paper No. 2023/169 (Washington, DC: International Monetary Fund, August 2023), https://www.imf.org/en/Publications/WP/Issues/2023/08/22/IMF-Fossil-Fuel-Subsidies-Data-2023-Update-537281.

As of this writing, more than 2000 coal-fired power plants remain active worldwide, producing more than 12 billion kilograms of carbon dioxide annually.[9]

Coal propelled me to live a different life than my grandparents and great-grandparents. I sit here today, a product of my grandfather's mining in the coalfields of Appalachia to move our family toward land ownership. With some buddies, my uncle owned and worked a pick mine in the 1950s: the larger companies were switching away from manual labor in favor of automation—the new machine was the continuous miner. One summer running the loaded mules in my uncle's mine cured my dad of ever going underground for permanent work, or, that is, as permanent as the boom and bust of coal demand was at the time. This same uncle had just finished a gig welding crude oil storage tanks in California. There was a time when someone might have understood these energy equivalents beyond theory—the capacity of a train or a truck still made sense to be measured by the strength of a horse. A technological work imaginary could include the pull of horses: steam and oil engines running on magic that you did not, in the same way, have to feed and care for, wipe down, steward, brush, take to pasture, or water. Coal and oil ran through the hands of men elsewhere, hidden from the centers of urban life in coal towns or oil fields, Big Rural places feeding societies' centers and rocket-growth by drilling. Horsepower filled our visions of constantly moving to new horizons, dragging forward our engines of progress.

I was raised on stories of my grandfather's and uncle's support for the United Mine Workers of America (UMWA). A major turning point in our family history was the New Deal and its legalization of union protections and negotiation, materializing the connection between my family's labor and the technological capacity of industry. I was raised on the stories of these men and their mastery of technology, mostly in rural places, to bring energy or food to the tables of people who no longer needed to sully their hands.

It was 30 years ago that my grandmother told me, with a bewildered look on her face, that that winter in these same coal-filled mountains, the ground had not frozen. Every winter of her life had been spent in the mountains of Appalachia or a valley near them in the Shenandoah. Then, at 78 years-old, she sat there with a distant look, uncertain about what this meant, when the weather and the mountains had always been one way, and in one season became something entirely different.

I originally re-drafted this chapter some seven weeks after the geologic event of Hurricane Helene. As of that writing, the frost in our region had still not happened. It was at least these same seven weeks late. In the region north of us, the Kanawha River Valley, flowers became confused, awakened in November by a brief relief of rain from the terrible drought that had plagued that region since May, making the humid hollers mimic the tinder of the California desert.

My husband's hometown was the main staging area for supplies and rescue for Hurricane Helene. For seven weeks after the event, as best as we could, either he or

[9] Global Energy Monitor. *Global Coal Plant Tracker*. April 2025. Accessed May 5, 2025. https://globalenergymonitor.org/projects/global-coal-plant-tracker/.

I headed into Western North Carolina with some kind of assistance for friends, or in my case, to try to take a tiny bit of weight off already terribly overburdened community-based organizations there. Also, my training in the nonprofit and NGO sector before I entered academia honed me to understand that there are always people who fall outside the official channels for distribution of assistance. There are always people with valid reasons for not going to get the aid where it is and needed it brought to a seemingly safer place for them instead.

My first post-Hurricane trip into Western North Carolina was to bring some culturally appropriate fresh food, and warm clothing, to a trailer park. Wedged into a steep hollow, now cut more like a narrow ravine, the trailers had been lifted by the force and deluge, and then sailed and stacked one behind each other. Two of the trailers closer to the road, while flooded, had not moved, and there now between them was a tent, where one person was living, and beside that, someone had brought and placed a small camper. About 15 people were still residing in this trailer park. Almost all of them were farm workers from Honduras and Mexico, living in the high mountains to assist with the Christmas tree harvest, the greenhouses there, and tobacco. Now, they had no running water of any sort, and the park's septic had also collapsed. I was told by my contact there that a Latin American person had gone into one of the local aid stations, and an out-of-state volunteer aid worker had grabbed the material assistance out of the immigrant's hands—that news of this incident had spread through the Latino community very quickly as it would, and, thus, from this incident and others, farm workers, Latinos, and other folks that felt and were treated as marginal did not feel safe, and likely were not safe, going to any of the official or established aid centers.

My second and longer time up to Spruce Pine and Burnsville, high up in Western North Carolina, the algorithms for Google Maps weren't working correctly, so they sent a lot of us by way of roads not actually open, or not open to non-locals. In any case, I was driving a 1990s model Suburban pulling a trailer of food on a road along what once used to be a flowing river.

There, lower mountains to the east and higher mountains to the west flank a mile-wide bottom, and, where the river had been, an estimated 200,000 years of usual geologic shift occurred only in 12 hours. All along where the water had leveled in the bottom for maybe 3/4 of that mile now sat boulders or large rocks, as if the bricks and mortar of the mountains themselves had collapsed. The bottom was filled for miles. Pools and puddles of water peeped from around the rubble. Part of the railroad in that valley had been cleared and repaired. One bulldozer and one piece of heavy equipment loaded boulders into a train car.

Back on the correct path and making it up a steep secondary road to the town of Spruce Pine, I assisted with an event put on by a Church of God from Morganton, North Carolina, where a lot of movers and shakers of that church were women originally from Latin America, now married to American men. They had organized assistance, advertising it in Spanish directly to the Latino community there. And we had a big turnout. Quite a few other non-Latino folks wandered through because they too reported not feeling safe going to the military-cosplaying aid stations. A young man, probably in his 20s, arms full of tattoos, face full of piercings, ears full of gauges, asked me if we had a blood pressure cuff available. I promised him that I would go get him one before I left town. He also needed allergy medication. I told him I knew what he meant, that although I live 2 1/2 hours away, since the hurricane, my allergies, too, had been very stirred up. He said that his had been awful, and that he just really needed some relief. He had lost two jobs since the hurricane started, but since he was handy, he had landed one being and doing maintenance at the hotel two doors down.

One rather shattered-looking middle-aged white man was intercepted by a Latin American woman volunteer and led to get a plate of warm food.

Two of my contacts fell into conversation and pointed out to each other and to me a particular person to avoid, an apartment building manager who was not letting residents receive any assistance being brought there. Residents had no access to uncontaminated water for showering or washing anything, and, she was not letting any clean water systems be set up.

Also, reports had been broadcast in English for several weeks that people should not touch the mud they were cleaning out of their homes because it had become infected not only with biological chemicals, but also with chemicals that had washed down through the river system from factories and paper mills along the valley bottoms. The day of this Spruce Pine event, we learned that many local farm worker communities had not gotten the message not to use creek water for washing their clothes or for bathing. They had known not to drink it, but they had not known not to even boil it and use it. So, their skin was peeling, their clothes were falling apart from having used it.[10]

During the aid event, the Church that organized it and other volunteers like me distributed meals, fresh food, hygiene items, household products, medicine we had hauled in. I followed through with a plan I had—there must be this kind of aid there locally that just needs to get redistributed that I, as a white woman, can go and get and bring to this community. I drove the U-Haul I had rented to one of the aid stations nearby. Online, it had looked like a large, stocked warehouse. Live, it was instead a rather nice wood frame house up on a hill, but the "warehouse" was actually an unused barn open on two sides. Some white guys were standing around on the drive to the house, cooking and eating barbecue. Another guy with a buzz haircut and wearing fatigues, not actual military, walked around as if on patrol, a weapon strapped to his side.

[10]The following week I arranged for someone I know through mutual aid connections to go up there and install an IBC tote for water for that complex, and the woman running it would not allow it to happen. My contact came away with the impression that there was potentially human trafficking or other kinds of prostitution happening in those apartments. Then we wondered if we were putting more people at risk by meddling there. After this, I focused later efforts on low-tech individualized water harvesting systems, where you catch your rainwater in one bucket, and then pour through a filter that is connected to the lid in another bucket. The issue had been with this, even, that after the Hurricane, the region had returned to drought. Then, came the promise of a wet, rather than snowy, winter. People needed individualized and discrete means of catching rainwater for washing and personal use.

As soon as I asked about stocking the truck to take elsewhere, these guys seemed suspicious of me. The two white women in charge insisted they had already made contact with me before. In the warehouse proper, there wasn't much of use to the people that we were distributing to: cat food and litter, uncooked pasta (what water were you supposed to cook that in?), a few small bags of rice, lots of junk food, some canned dog food. I did not announce what community I was bringing this aid to. I tried another official aid station, and it really didn't have much other than old and used clothing. The takeaway from this was that a lot of the aid was getting redistributed, however, the idea of these places feeling very militarized was also accurate.

In the end, I ended up transporting a load of diapers and leftover baby items to a warehouse for later day-to-day distribution, and then I went up to the only retail store functioning, the Walmart. In the parking lot, an animal rescue organization had set up some trailers. A very militarized-feeling volunteer group had set up their stations of campers and a fiberglass yurt. I ran into some religious-affiliated volunteers from West Virginia up there to use the shower station. I walked into Walmart and although it was open, you could not go to the restrooms.

I found a suitable blood pressure monitor and some allergy medicine for the guy with the gauges in his ears, and I thought to myself, if I were originally from a militarized dictatorship, or a presidential candidate had said that as a migrant, I was taking aid away from other local people, I certainly would not go anywhere near this Walmart.[11]

11.1 We Are All the Coalfields Now

In the 1960s, one of the biggest fears of the American population was potentially losing their jobs to automation.[12] Several years agoa preacher in the Pocahontas Coalfield of West Virginia and Southwest Virginia penned a thought piece about this fear in our current age, and the coming of vastly more automation. In "I Have Seen the Future and it Looks Like Appalachia," he explains that the current problems (poverty, addiction, lack of entrepreneurship or investment, etc.) in the coalfields started with the introduction of the machine the continuous miner, which reduced the need for a human workforce in the coalfields by half by the end of the 1950s. This led to mass exodus and all the subsequent woes of depopulation.[13]

I, too, have seen a future that looks like Appalachia, our future everywhere as this past/present future of Western North Carolina. Of every valley there, I now wonder

[11] Alison Durkee, "Hurricane Helene: Trump Falsely Claims FEMA Money Going To Migrants," *Forbes*, October 4, 2024, https://www.forbes.com/sites/alisondurkee/2024/10/04/trump-fema-claim-debunked-agency-not-running-out-of-money-because-of-migrants/.

[12] Thomas, Jerry Bruce. *An Appalachian Reawakening: West Virginia and the Perils of the New Machine Age, 1945–1972*. Morgantown: West Virginia University Press, 2017.

[13] Lowe, Travis. "I've Seen the Future. It Looks Like Appalachia." *100 Days in Appalachia*, April 2, 2017. https://www.100daysinappalachia.com/2017/04/ive-seen-future-looks-like-appalachia/.

if 40 centimeters of rain will soon fall nearly all at once and propel even more of the rocks above onto the valleys and fields below. I also wonder now, with the deliberate State destruction of the National Oceanic and Atmospheric Administration and climate science, will there be any tangible and effective warning? With the dismantling of the Federal Emergency Management Agency (FEMA) will there be any assistance?

But let me continue with some more recent rural stories, either my own or recounted directly to me.

Directly after Hurricane Helene, local, state, and federal actors worked to assess the damage and connect people with resources. Throughout social media, people posted again and again that the Democrats had a weather machine that had manufactured this horrifying event. Other AI-pushed information was that government aid was not helping anyone, and that Federal Emergency Management Agency aid was to be denied. FEMA workers in the high mountains (where I later assisted bringing aid to farm workers) were threatened by people wielding guns, so the federal government withdrew its workers.

Scholar Pippa Norris illustrated nearly 25 years ago in *Digital Divide: Civic Engagement, Information Poverty, and the Internet Worldwide*, the split in access to accurate information between the haves and have-nots.[14] The difference now being that a responsive artificial intelligence can tailor to our prejudices, manufacture seemingly "real" people and responses on the fly, fill standard and niche social media with impoverished information or anemic or unproven sources, and authenticate nonsense by circling the nonsense with the sheer power of numbers of real "likes" or manufactured but seemingly "real" comments in agreement.

In 2022, my father passed, and I decided the following summer of 2023 to take a break and go see friends in Europe I had not seen in 20 years.

In an Uber in Berlin, a very amiable driver originally from Tunisia chatted with us in a partially Scottish brogue, having recently relocated from Edinburgh. We chatted about our owning a small farm in North Carolina, and he immediately livened up, as he had grown up on a very rural oasis with his grandfather. This oasis had been in the care of his family for hundreds of years. His grandfather went out every day and took care of the many perennials and trees. His grandfather had names, the kind of names you reserve for people, for these trees. This palm was so and so. This peach tree was so and so. Two years prior, the oasis had begun to get so hot that the trees that provided the livelihoods for his family and the other villagers began to die. This cabby from Tunisia explained to us that his area of the world had no factories. They had no coal fired power plants. They barely had any cars. Yet, here was his family's farm, dying from heat. And, in response, his grandfather grew so despondent that he too died of a heart attack.

In Honolulu in November the same year, I put on a performance during the Making and Doing section of the Society for the Social Studies of Science (4S).

[14] Norris, Pippa. *Digital Divide: Civic Engagement, Information Poverty, and the Internet Worldwide*. Cambridge: Cambridge University Press, 2001.

I had intended to lament and grieve the losses experienced in single sector rural places. I was starting with my own, and I brought coal to smash; I invited others to join me and talk with me about coal-fired power plants and coal-producing places.

And as I sat during my "performance," the questions from 4S scholars and students shocked me. Some thought that if they touched the coal, they might get contaminated, that the little bit of coal dust that we were creating from smashing coal might give them black lung or make them sick. Some amazed to me that I had been allowed to bring the coal to Hawaii in my suitcase. I patiently explained that coal did not function like uranium, that this small amount was not poisonous to the touch and very unlikely to do any damage to their lungs. I wondered at their fear of, and my familiarity with, this society-central material. I felt alien, I feel alien, that what I know as central remains invisible. Large Technical Systems indeed. Materiality for sure. Ah, Langdon Winner.

I was glad to have connected and shared, but I was despondent in the halls during those days filled with scholarship on artificial intelligence, and nothing on mining, mining engineering, much less coal, and very little about any sort of actual rural resource extraction, production, or even about manufacturing. But I assume that since then, that there has become an STS interest in mining after all vis a vis "critical" minerals. Latour might as well have been on the moon as at the Ecole des Mines de Paris. Maybe the next Making and Doing I submit ought to be my sucking on coal like miners did for indigestion or feeding coal to hogs that old timers swear eat it like candy.

11.2 Ephemeral Steam of a Coal Plant. Fickle Science and Technology Studies Algorithm

Off-shore places: The natural-resource based society remains central with largely rurally-produced resources affording the metropole the luxury of focusing on and managing knowledge production. Given the continued population growth of the world and the destruction of the environment and our lives to sustain growth economies to support this population expansion, this wealth transfer from rural or off-shore to metropole continues to be black-boxed, invisible, making the raw material resources and the people involved in their extraction or production also invisible, disenfranchised. The invisibility of Big Rural abounds.

11.3 The Coalfields Are Also Western North Carolina

As of this writing, more than 80% of the debris from Hurricane Helene remains unmoved, unmitigated. The Norwegian writer Jens Bjorneboe wrote a three-part trilogy that was the pinnacle of satire, the heart of which circled on World War I and the killing fields of the Battle of Verdun. In it, he estimated the items that were left behind to create compost, enriching the soil. Much of the debris of Western North Carolina will remain where it is until the next similar geologic event knocks it loose.[15]

A "climate flickering" is a rotation between extreme drought and heat and extreme rain, which can indicate the climate of a place may soon radically change. With Hurricane Helene, this is what happened in Western North Carolina, and then also happened along the same latitude a few weeks later in Valencia, Spain. Would that Hurricane Helene worked upon us as did the Lisbon earthquake of 1755 on its rulers—creating for our government a new assessment of how we ought to build and create—a new era of enlightenment, of science, of planning.[16] If only it pushed us to begin planning for the much larger disaster that this hurricane climate-flickered.[17]

Climate change is "un" black-boxing.

With Hurricane Helene, 12 tornadoes occurred around Mount Mitchell, and 23 inches of rain fell on one side of it. In human memory, there had never been tornadoes there. We live in a major laboratory experience-experiment. All of Western North Carolina became "unwitting subjects for a vast series of biological and social experiments." On that note, also never having happened in human memory, recently in the high desert in Saudi Arabia, it snowed. Compared to what actually is happening—the earth's death by 100 million cuts, the collapse of regions, the collapse of insects, the collapse of wildlife, the collapse of "world order" and the order we understood and the classifications we catalog, the twentieth century conceptualization of full-on nuclear destruction or of Biblical apocalypse seems merciful.[18]

To quote Winner:

> "Where does my own personal interest in this topic come from?" (166) "Why have I chosen to approach them in the way that I have?" (166) "To talk about these experiences directly may help reveal the ideas and sensibilities we hold in common." (167)

Fall 2024 I was approached to participate in a potential National Science Foundation grant around communicating extreme weather, and the meteorologist principal investigators echoed to me what I had said to some colleagues earlier that week, *with Hurricane Helene we didn't have the models for what had happened.* The meteorologists assured me that they had a good handle on what extreme heat and cold produce, but they didn't have any models that included a weather event that had

---

[15] Bjørneboe, Jens. *Moment of Freedom*. Translated by Esther Greenleaf Mürer. New York: Norvik Press, 1999.

[16] Dynes, Russell R. "The Lisbon earthquake in 1755: the first modern disaster." (2003).

[17] Dakos, Vasilis, Egbert H. van Nes, and Marten Scheffer. "Flickering as an early warning signal." *Theoretical ecology* 6 (2013): 309–317.

[18] Kinzer, Stephen. *Poisoner in Chief: Sidney Gottlieb and the CIA Search for Mind Control*. New York: Henry Holt and Company, 2019.

Winner, Langdon. *The Whale and the Reactor: A Search for Limits in an Age of High Technology*. 2nd ed. Updated edition. Chicago: University of Chicago Press, 2020.

simultaneously included a hurricane, a deluge, tornadoes, landslides, and floods, right after extended drought. They also assured me we would be just fine for an application in the then-upcoming administration; they had all weathered the last one just fine.

For my day job, I've been working for several years on investigating means of preparing farmers for climate change. Until recently, the research site for the team I lead was Northeastern North Carolina, which gets hit by hurricanes every few years. Until the spring of 2025 when the federal agency sponsor of this research sent a memo that grants could not focus specifically on any one demographic, we engaged specifically with Black farmers.

In December 2024, I sat down with a colleague of mine, a soil scientist, who got caught in Hurricane Helene in Asheville. He and I agreed that no amount of climate mitigation suggested practices would have helped any of the farmers where the mountains and boulders rolled down onto their farms.

Hurricane Helene is long out of the news cycle. There will be no Lisbon earthquake style repercussions. The ineffective ruler with dalliances and caprice will not retire to his tent and let an interested and competent intermediary take his place.

In the Ohio River Valley where I lived for some time as a middle and high schooler, often in summer it was so hot and humid that by late afternoon, condensation had formed on the outside of our brick house. During the same fall as Hurricane Helene, for the first time in human memory, the Ohio River Valley, directly north of Hurricane Helene's path, suffered a drought that threatened the area with exploding into Western-US-style wildfires.

The drought lasted through the winter and on until 2025. As of this writing, that region is still in drought.

In February 2025, I attended for the first and last time the National Oceanic and Atmospheric Administration's Extreme Weather Social Science Summit held at the National Conservation Training Center in Shepherdstown, West Virginia. If you have never been, and I am not sure you may now ever get to go, it is or was a gorgeous facility designed like a lodge, and featured feats and luminaries in US Fish and Wildlife service and in the Conservation movement and Service.

For the summit, I dropped into sessions where participants went about their usual business of roundtables, working groups, discussions, and seminars. I remained amazed that the conference proceeded with little discussion of the current administration. By contrast, I felt like maybe we walked around a bit too Weimarian, certain that what seemed to be coming couldn't actually reasonably, rationally really happen.

The first full day at lunch I found myself drawing an audience as I foretold what I was afraid was about to occur, and that we might all need to be thinking through other employment options, much less other ways to accomplish our work. I imagined my social science colleagues imagined me rather like a young Germaine Greer about to burst into the lunch hall at Cambridge and yell about the patriarchy. What I was saying was true, but no one there wanted to hear it.

During lunch on the second day, I found myself chatting directly with an economist originally from Bosnia, who had listened in the day prior. He had already lost his long-term contract job with a Beltway agency that had assisted developing countries but had been defunded. That second day he and I worked our way through all the occurring and coming scenarios. We of course underestimated the deliberate massacre about to happen to the Fifth Branch and US State-sponsored expertise.

Before we parted, he looked at his lap, then turned to me, his mouth twitching, "I don't think I can bear it. I don't think I can bear to lose another country, to have another country break apart."

What am I supposed to say now about rural places? What kind of prescription am I supposed to make in theory or language or conceptualization for their current state or where they're going?

What am I supposed to say about migrant farmworkers, cowering and afraid to go get aid in a rural place because a presidential candidate—now a president—demonized them like we used to demonize the whale? And now, he is deporting migrants to a prison in El Salvador. What am I supposed to do?[19]

What I considered earlier in 2024 for this chapter doesn't matter now.

I had thought some about how maybe, over the last 150 years, the totalitarian workplace had trained us to accept totalitarianism in other facets of our lives. I had considered what historian Timothy Mitchell outlined in his book *Carbon Democracy*, where the material facts of coal miners being able to lay down their lives for coal production 120 years ago led to getting the union, to the practice of physical and active democracy.

I knew 4 days before the 2024 election that president 45 would win again. During the first presidency, I was minding my business in our diesel Suburban pulling hay, and an older man in a mini-truck covered in Trump stickers (our vehicle had no stickers) hit my trailer on purpose. Everyone at the place where we were saw it and assisted with mitigation—ushering the man along with his food, giving the woman in her car behind him free food, and discounting mine.

In early 2020 I had someone look up my political affiliation online when I was searching the county for our lost Livestock guardian dog, and when this man saw what my affiliation was, he started a disinformation campaign about me online that I had hit HIS dog, and I had to get detectives involved and threaten a civil lawsuit.

Five days before the 2024 election my kid and I were at a large street festival in our local town, waiting for his turn to do a public Irish dance performance, and a tall, younger man shoved me with both hands really hard out of the way and said, "I'm just making way for my daughter to get through." The act stunned me, so I sort of bounced back to my place, so he did that to me again. Right then and there, I knew in my body. I was like, oh, yeah, I remember this. I knew what was coming on the following Tuesday.

From 2010 to 2014 I devoted my life to understanding single sector economies, rural places, and development.

There is nowhere to hide from climate change.

There is no longer any refuge in the rural.

[19] Dale, Daniel. "Fact Check: Trump's False Claims About Hurricane Helene Response in North Carolina." *CNN*, October 6, 2024. https://www.cnn.com/2024/10/06/politics/fact-check-trump-helene-response-north-carolina/index.html.

Apocalypse smells like: *The earth has shifted underneath our feet. The water now has a sulfur smell. We stand on stones rather than on fertile bottom land.*

I want to travel to Langdon Winner's past to join him in contemplation on the beach. I want us to issue a joint apology to that whale.

We missed the mark, Whale. We missed where our systems were fragile. We missed where and how the totalitarian impulse travels and how fast. We missed what is going to destroy us all. Whale, we missed what is going to destroy you. Whale, we missed what is destroying you.

And you, yes, you, Whale, that very whale, maybe still swim near that beach. Maybe you are now choking in that same water.

Langdon Winner, we have heeded very little.

We have, in our socio-technical renderings, offered no harbor to any being.

Chapter 12
Technological Politics and the Rhetorical Situation

Christopher Lee Adamczyk

Keywords Rhetoric of technology · Rhetorical situation · Technological politics · Deliberation

I

One of my grandfathers was a reliability engineer for GM. The other was an auto mechanic. I remember them being together a few times, usually when they visited my parents for special occasions. During one of these occasions, my parents' car needed a bit of maintenance. Wanting to help, my grandfathers set to work on what ostensibly were simple tasks. Tune this up. Rotate that. Flush something else. Halfway through their work, I remember hearing a brewing disagreement. The engineer was upset with the mechanic for complaining about the design of the engine compartment—how he couldn't reach important bolts and how everything was crammed into tight spaces. For him, these were design decisions that ensured new vehicles were more efficient than what GM had produced in the past. The mechanic was incredulous about any claim that designs that made vehicles difficult to repair improved the vehicle. For him, a car's most useful trait was ease of maintenance and symmetry with the past. Why reinvent the wheel when it still rolled? The debate was never truly resolved.

Their disagreement became a flashbulb memory for me that took on new life when I began graduate study at Rensselaer Polytechnic Institute. Studying with Langdon Winner further called me to consider the dynamics of my grandfathers' dispute. The more I learned, the less my reflections centered on familial belligerents and the more they centered on the situation itself: One man an engineer, the other a mechanic. An argument over the design of an engine. No real resolution. An

C. L. Adamczyk (✉)
Farmingdale State College, Farmingdale, NY, USA
e-mail: adamczc@farmingdale.edu

© The Author(s), under exclusive license to Springer Nature Switzerland AG 2026

A. Fox, D. Cressman (eds.), *Arrangements of Power: Tracing Langdon Winner's Legacy Within and Beyond the Philosophy of Technology*, Philosophy of Engineering and Technology 50, https://doi.org/10.1007/978-3-032-17371-3_12

unchanged but now functional car. Lay outrage at what was perceived by the other as logical innovation. This one memory embodied all the themes I encountered in my readings about the relationship between technology and society. But rather than an account of a distant historical event, it was one I had been present at myself. And, it was little more than an average event. Laid bare before my eyes was a lucid example of technological change and the interpersonal friction that accompanies it.

As luck would have it, the element of my grandfathers' dispute that stood out to me most—the argument—became the focus of my studies. While my work is eclectic, it is bound to rhetoric and a desire to understand the public debate about technology. What we say about technology and how we say it matters because it is a vital avenue through which we can shape technologies. It illustrates our beliefs about technological change, where we see the boundaries of reasonable action, how our communicative practices regarding technology are enmeshed with the demands of devices themselves, and the broader political implications of technological systems. And to be colloquial, boy, howdy do we talk a lot about technology nowadays.

In this chapter, I consider Langdon Winner's characterization of high technology society and examine how it intersects with rhetorical insights about public discourse. My thoughts are guided by Winner's observation that "the technical equivalent of the Archimedean point—a place to put the lever so that one can move the mechanism—is often missing" (Winner, 1977). In other words, because the sociotechnical systems that have emerged during the twentieth century characteristically are large and complex, our ability to leverage control over them is significantly eroded (Winner, 1975). Robust public and private debate about technology feels futile when emergent technologies slowly develop the already critiqued traits present in older models (Winner, 1992). My grandfathers' argument reflects this reality. Robust debate. New models of cars. A clear trend toward incomprehensible complexity is defended with cries of efficiency. But efficiency for who and what? Nevertheless, we know that discourse can exercise some degree of influence over technology—it's the very reason my grandfathers debated. What, then, explains the lack of influence that public critique now seems to leverage? Why does it not seem to get much done anymore? Where has the Archimedean point gone? Such questions are of vital importance to both technology studies and rhetorical studies.

I proceed in three parts. First, I identify resonances and commonalities between Winner's work and rhetorical studies. Then, I consider how Winner's work has influenced rhetorical scholarship, with a focus on the work of Carolyn Miller. Lastly, in light of Winner and Miller, I consider rhetoric a situational phenomenon, arguing that the material rhetoric embodied in technical artifacts calls us to reconsider the nature of rhetorical situations concerning technology writ large.

II

While the significance of deliberation about technology is not lost on Winner, rhetorical studies uniquely resonate with the thrust of his work while offering definite tools for analyzing discourse about technology. While traditional theories of rhetoric have centered on written and spoken discourses, more recent trends have considered non-symbolic communication and spawned subfields that consider material rhetoric, ambient rhetoric, and affective rhetoric (Barnett & Boyle, 2016). These trends in rhetorical studies mirror concerns Winner presented throughout the corpus of his work. Winner's articulation of the theory of technological politics juxtaposes the shaping of technology by human discourse with the material and structural demands placed on humans by large sociotechnical systems. The result is a depiction of contemporary society within which humans possess the power to construct technology in any plausible form yet repeatedly choose to develop new technologies that are symmetrical with extant, often authoritarian, systems. Winner's observations are reflected in my grandfathers' argument over automotive design trends that seem to serve efficiency over human control.

Winner labels this phenomenon *technological somnambulism*, or the tendency of humans to "sleepwalk through the process of reconstituting the conditions of human existence," and offers something like deliberative rhetoric as a counterbalance (Winner, 1986). In *Autonomous Technology*, Winner suggests that a solution to the unchecked growth of technical logic is democratic participation in innovation by all stakeholders. The goal of participation would be to develop technologies that have designs "immediately intelligible to non-experts" and to critique emergent technologies on "the degree of dependency they tend to foster" (Winner, 1977, pp. 326–327). In *The Whale and the Reactor*, Winner relatedly contends that our abandonment of technological assessment to the logic of technology invites "spiritual rigor mortis" (Winner, 1986, pp. 175–176). By this phrase, he means that we have undermined our humanity by abandoning rigorous deliberation solely for risk assessment and other quantitative-oriented methods of judgment. For Winner, while it is certainly possible to bend technology to the will of humanity, it requires immense dedication to democratic deliberation and development and political systems amenable to direct action. Even when this is accomplished, technological systems will still, in a sense, argue back. Bluntly, even if my grandfathers' had solved their disagreement and proposed a compromise, GM would have ignored it if it ran contrary to the immediate needs of their system. As Winner argues, there is a paucity of citizen "opportunities for direct... technology policy-making," and where they do emerge, they stall out on "futile rituals of expert advice" or "interminable disagreements about which technological choices are morally justified" (Winner, 1992, pp. 350, 352).

Winner looks to strong democratic models of technological development as a solution to technological somnambulism. In strong democratic models, stakeholders have greater say in the design and adoption of technology and, thus, distinct ways to voice concerns about the political dimensions of technological design (Sclove, 1995). Essentially, Winner calls for the development of a practical rhetoric that empowers citizens to make impactful, knowledgeable technological choices *and* acknowledges the impediment to efficacious deliberation that robust, ossified technological systems present. And, certainly, rhetoric can deliver—at least theoretically. Especially important here is rhetoric's non-reliance on a rational choice model of political deliberation, with more emphasis placed upon "relations of mutual dependency and the role of trust in deliberative processes" (Hauser & Benoit-Barne, 2002). From this vantage point on deliberation, everyday practices "reflect values, aspirations, concerns, affiliations, boundaries of acceptance, levels of tolerance, and the like that shape our civil judgments" that ultimately "aid our capacity to assess risk and form expectations about others that can guide our actions and contribute to a civil society in which trust becomes a public good" (Hauser & Benoit-Barne, 2002, p. 272). The endpoint of cultivating critical rhetorical practices centered upon community deliberation is to foster individuals able and willing to engage in public discourse that directly transforms materials and social conditions (Lyon, 2013). This goal is relevant to Winner's concerns about the erosion of citizen engagement in democratic practices and how that erosion has an outsized influence on technical logic.

Some may be tempted to see rhetorical studies as a roundabout way of proving the social construction of technology. Winner's critique of purely social constructionist approaches to understanding sociotechnical contexts is relevant regarding the temptation. In "Upon Opening the Black Box and Finding It Empty," Winner acknowledges the worth of social constructivist approaches while cautioning that they cannot offer answers to all questions about technology nor offer firm footing for cultivating public engagement with technological development. Of special concern for Winner are the approach's "disregard for the social consequences of technical choice," reliance upon "relevant social actors" in its analyses, sidestepping of "dynamics evident in technological change beyond those revealed by studying the immediate needs, interests, problems, and solutions of specific groups and social actors," and "apparent disdain for anything resembling an evaluative stance or any particular moral or political principles that might help people judge the possibilities that technologies represent" (Winner, 1993). These critiques stem from Winner's deep concern with the potential political character of technological design, especially when artifacts imbued with such a political nature have become more or less ossified.

In contrast to purely social constructionist interpretations of technological development, rhetoric has long displayed an interest in politics and morals, both regarding technology and otherwise. Rhetoric is integral to politics because politics involves making decisions in situations where collective agreement is elusive. Rhetors make their cases about particular decisions using a variety of claims that reflect epistemological, moral, and material understandings of the past, present, and future. Understanding that talk and argument contribute to collective decision- making and identifying those engaged in it is not enough. One must also understand past and present decisions, the claims that led to them, and how ideological interpretations influence the range and potential success of rhetoric in the present. Moreover, the power dynamics between stakeholders must be considered. And, perhaps most importantly, Aristotle's definition of rhetoric—the ability to identify the available means of persuasion in any situation—leaves open the understanding that any number of people, events, or (from material rhetoric's perspective) objects possess suasive affordances. We cannot forget that technological devices, through their design and aesthetics, produce human reactions that are definitively rhetorical (Brummet, 1999). Such reactions can result from an encounter with a singular technology or, as is perhaps more common, encounters with a "system of objects that generates meaning" (Brummet, 1999, 27). More broadly, Scot Barnet and Casy Boyle argue that humans and objects are co-constitutive and that things exist in "vibrant and rhetorical ways" (Barnett & Boyle, 2016, p. 6). Thus, it is disingenuous to characterize rhetoric and discourse as primarily constitutive of technology without accepting that the relationship between people and technology is a two-way street. It is ultimately the rhetorical critic's job to identify such persuasive affordances wherever they are and to determine where they exercise influence, where they do not, and why.

III

Despite considering similar tensions between humans and their material environment, rhetorical scholarship directly related to Winner's work is limited. Rhetorical studies generally were slow to uptake ideas from emergent STS scholars—a state of affairs amplified by differing trends in each field during the 1970s–1990s. However, despite this lack of robust cross-pollination, it is not uncommon to find Winner or other authors associated with the theory of technological politics cited in rhetorical subfields concerned with technology and especially in work engaging with datafication, emergent media technologies, and materiality broadly.

The most robust example of Winner's influence on rhetorical studies is in the work of Carolyn Miller. Miller's early work is keenly sensitive to the theory of technological politics, considering how rhetorical terminology can contribute to our understanding of technological development and raising questions about how rhetoric is enmeshed with large sociotechnical systems. For example, in her essay titled "Opportunity, Opportunism, and Progress: Kairos in the Rhetoric of Technology," Miller draws upon the rhetorical concept of *kairos*, or timing, arguing that rhetorical characterizations latent in technological forecasting are a vital part of how technologies develop down certain paths (Miller, 1994). Miller argues that the construal of kairos in technical discourses "functions primarily to create opportunities for opportunity" by making the emergence of particular technologies feel inevitable and thus whetting public appetite for them (Miller, 1994, p. 93).

For Miller, understanding rhetoric's influence on technological development in this manner is necessary to address political uncertainty associated with technological adoption. Miller's viewpoint aligns with Winner's observation that control over technological development allows those with a vested interest in current political and social ways of organization to foreclose upon the unforeseen social and political consequences new technologies bring with them (Winner, 1986, p. 25). Because technological predictions are typically produced by powerful stakeholders—governments, corporations, academia—with a vested interest in political environments that are symmetrical with their predictions or the technologies they develop, public expectations about the future stemming from those predictions prepare them for more or the same more often than not.

Before moving on, it is worth noting that Miller's observations on kairos do not suppose that rhetoric socially constructs technology so much as they suggest that rhetorical constructions enable technological systems to grow more or less by their own logic, especially when corporate or governmental groups produce them.

Miller's essay titled "Technology as a Form of Consciousness: A Study of Contemporary Ethos" more firmly makes this case. Here, Miller addresses *ethos*, or the characteristics that enable an individual to be perceived as trustworthy, in a high technology society. She argues that the ethos that lubricates trust in technological society "must necessarily be technological" and that "the actions we engage in, their purposes, and the conceptual habits they encourage delineate the consciousness which formulates our discourse" (Miller, 1978). If our ability to construe ourselves as ethically trustworthy is the crux of ethos, then a society in which high technology is foundational to stability engenders trust in discourse that supports or strengthens technical logic. For better or worse, we find rhetoric associated with the status quo trustworthy. As Miller suggests, "the ideology surrounding high-context technology has prevented the achievement of communal understandings of prudence and efficiency," instead erring to definitions most in line with technical logics themselves (Miller, 1978, p. 236).

Although it is unclear if Miller's essay addressing ethos and technology was directly influenced by Winner's work—Miller's essay was published in 1978, only a year after *Autonomous Technology*—it heavily draws upon theorists like Jaques Ellul and John Kenneth Galbraith whom Winner also uses to craft his perception of the theory of technological politics. In fact, Miller's work on ethos bears a striking resemblance to Winner's essay "Techne and Politeia" in which he argues that the rationality of modern technical systems has itself become a form of political organization (Winner, 1986, p. 40–58). When considering Miller's arguments concerning kairos, we must keep both her observations about ethos and Winner's observations about the relationship between technics and politeia in mind; no one ever engages in discourse from a neutral standpoint, only one colored by the context within which they live.

While Miller's work on kairos and ethos put rhetoric in conversation with the theory of technological politics, her essay "Learning from History: World War II and the Culture of High Technology" offers novel terminology that further characterizes their intersection. Winner's influence on this terminology is marked. Undergirding Miller's analysis is Winner's claims that citizens in high technology societies lack in-depth knowledge about technologies that enable their way of life and that large sociotechnical systems develop in autonomous ways as a result (Miller, 1998). Emerging from a discussion in which she addresses post-second world war attitudes toward technology and their bearing on the distinction between the rhetoric of science and the rhetoric of technology, Miller raises questions about how certain technologies are widely propagated and how they come to acquire cultural power (Miller, 1998, p. 307). To understand rhetoric's role in this process, she suggests three categories of discourse concerning technology: rhetoric *about* technology, rhetoric *within* technology, and rhetoric *from* technology. Under this paradigm, rhetoric about technology refers to "public representations or debates in public policy forums," rhetoric within technology refers to "private, proprietary discourse by which technological work gets done," and rhetoric from technology refers to "the ways in which values and thought patterns developed by technological work extend to and pervade other cultural arenas" (Miller, 1998, p. 307). Miller's work on kairos and technology slots well into the categories of rhetoric about and rhetoric within technology, depending upon who does the forecasting and in what context that forecasting is done. Beyond this, the categories represent a significant expansion of how we might conceive the relationship between rhetoric and technology. Together, they neatly suggest distinct ways in which persuasion and technology intersect, especially in traditional public discourse, technical communication within corporate, scientific, or governmental settings, and because of the material design of technology itself. In this tripartite interpretation of rhetoric's intersection with technology, we see reflected the concerns of speech rhetoric, material rhetoric, the social construction of technology, and the theory of political technology insofar that it identifies the influence of publics, proprietary organizations, and the material environment on the development, implementation, and use of technological systems.

Of Miller's three categories, rhetoric from technology most formally addresses Winner's concerns about the political nature of technological design. Critically, this category admits that technological artifacts possess something like their own persuasive capabilities stemming from their design. We can understand these characteristics as a type of *affordance* or the "relationship between the properties of an object and the capabilities of the agent that determine just how the object could possibly be used" (Barnett & Boyle, 2016, p. 10). Put simply, a well-designed object will invite a user to use it symmetrically with the logic that guided its design, although not require it. Miller reemphasizes this viewpoint in the forward of *Rhetorics and Technologies: New Directions in Writing and Communication*. Here, Miller writes, "technology, like rhetoric, can both push and pull us... by requiring us to do certain things in certain ways" (Miller, 2010). She colloquially refers to this dynamic as the "pushmi-pullyu," suggesting that "if rhetoric is the art that adjusts ideas to people and people to ideas, we might characterize technology as the art that accommodates the material world to people and people to the material world" (Miller, 2010, x). Whatever we name the phenomenon, Miller's observation that values and thought patterns are designed into technology and exuded through the use of technology is in line with Winner's examples in "Do Artifacts Have Politics" and "Techne and Politeia". Whether it is an ancient Greek galley or low bridges on Long Island parkways, some objects or large sociotechnical systems are simply more efficient, more usable, and more socially acceptable when we adopt the political viewpoints associated with their creation: to see this as a form of persuasion is no large leap.

Miller's work offers a unique example of how Winner and the theory of technological politics intersect with rhetorical studies. Importantly, Millers work provides a foundation from which we might consider rhetoric and its role in high technology society. As Winner contends throughout the corpus of his work, deliberate and thoughtful democratic forums are needed to counterbalance the momentum of technical logic. Rhetoric, be it employed for morally praiseworthy or morally base ends, is integral to the democratic process. We would be remiss not to raise questions concerning its form in high technology societies, as this directly intersects with concerns about our ability to democratically deliberate about technological design and technological politics.

IV
If rhetoric can embody and influence technical logic, how can rhetorical studies offer stakeholders guidelines toward democratically potent discourse about design? I address this question by putting Winner's and Miller's observations about rhetoric, democratic design, and the theory of technological politics in conversation with Llyod Bitzer's, Richard Vatz's, and Scott Cosigny's theories that address the situational nature of rhetoric.

12.1 Bitzer and Vatz: Determinism and Social Constructionism in the Rhetoric of Technology

The notion of a rhetorical situation was first proposed by Llyod Bitzer in 1968. From Bitzer's perspective, discrete discourses arise "because of some specific condition or situation which invites utterance" (Bitzer, 5). Bitzer's primary characterization of rhetorical situations is threefold: (1) discourse comes into existence necessarily as a response to an *exigence* or a perceived imperfection in reality marked by a perceived urgent need to address that imperfection; (2) many rhetorical situations arise without being adequately addressed by skilled rhetoric; and (3) the character of a situation exercises control over what is considered appropriate argument about it in the same manner that a question influences the character of the answer that addresses it (Bitzer, 5–6). Put simply, if we consider the cultural, political, and social timbre of any given situation, we find that the situation itself determines certain arguments, modes of persuasion, and other forms of communication. Thus, for example, Bitzer would contend that my grandfathers' argument about automotive design was not only called forth by the presence of a car that needed maintenance but also that the situation in itself suggests the most effective persuasive strategies for those arguments—*if and only if my grandfathers' could identify them.*

Like Winner's observation that the timbre of our highly technological society impacts the human ability to act freely, Bitzer's understanding of rhetoric emphasizes the influence of context on human action. Bitzer acknowledges that the most efficacious rhetorical actions are symmetrical with context. The rhetorical exigence in Winner's characterization of high technology society, for example, might be understood as "the erosion of human autonomy" or "non-democratic technological innovation." Conversely, if we were to ask Bitzer how citizens might democratically engage in discourse about technology, he would probably admit that doing so in a context dominated by high technology would be difficult.

Relevant to Winner's concern with motivating the public to participate in deliberation about design, Bitzer further identifies *audience* and *constraints* as integral to rhetorical situations. These concepts refer respectively to the people at which a given rhetorical message is targeted and the limitations imposed on discourses by the "persons, events, objects, and relations which are part of the situation" that "have the power to constrain decision and action needed to modify the exigence." Thus, for example, my grandfathers' arguments concerning automotive design may have been ineffective simply because they did not fully consider the epistemological beliefs of their sole audience that constrain how they interpret the world, be they commitments to the purpose of design, the nature of their employment, or even their beliefs about the value of the technological society.

On the one hand, and acknowledging Winner's concern for citizen participation in technological regimes, the target audiences of discourses stemming from exigences concerned about technology might be public stakeholders, engineers, or even governmental regulators. These audiences, drawn out, form the backbone of Miller's rhetoric of technology categorizations, each inviting definite ways of approaching technical rhetoric. On the other hand, constraints might include any number of phenomena, including, but not limited to, beliefs, attitudes, documents, facts, traditions, images, interests, and motives that intersect with being in a technological society. For example, Miller's observations on ethos and technology suggest the need to address constraints deriving from technical logic when composing arguments concerned with life in technological society. Considering the situation Winner describes in terms of rhetorical constraints is perhaps the most interesting and enlightening inroad, as Winner's body of work identifies various constraints that influence human decision-making about technology, including but not limited to techno-utopianism, apraxia, and complexity. More colloquially, if my mechanic grandfather had learned to offer his critique in the language of engineers and addressed the assumptions therein, it may have been acknowledged as a worthwhile viewpoint by my engineer grandfather.

If we fully apply Bitzer's logic to the situation of modern technology that Winner describes, then we must conclude that no rhetor has indeed arrived who can accurately read the situation and act upon it despite the acknowledged presence of beckoning exigence. This absence is a troubling prospect for certain and one that is difficult to accept despite the usefulness of Bitzer's terminology. However, the observation that rhetorical discourse is efficacious when symmetrical with its context does provide additional clarity to Winner's writings on risk assessment, decentralization, and reverse adaptation, which all point toward critique spoken in the language of technology. When juxtaposed with the theory of technological politics, Bitzer perhaps teaches us that the technological society muffles voices of critique not offered in the language of efficiency. The question then becomes how, or whether, it is easily possible to leverage real change in such terms.

Although Bitzer's characterization of the rhetorical situation identifies key aspects of the phenomena that are utilizable, it embodies a tendency to overlook the influence of rhetorical skill in and of itself. Moreover, if we are to identify rhetorical strategies suitable for leveraging change in high technology society, rhetorical skill in tangible situations is a tool for which we must account. More directly, can rhetorical skill shatter the need to offer critique in terms symmetrical with the language of high technology society? This provocation is born out in examples from Winner's work. For instance, in Winner's discussion of the appropriate technology movement found in *The Whale and the Reactor*, Winner makes it clear that social activists of the 60s and 70s were able to make lucid arguments for rethinking technological design, especially when supported by the work of theorists like Lewis Mumford, Jacques Ellul, and Herbert Marcuse (Winner, 1986, p. 66). Yet, these discourses existed apart from the ossified assumptions of mid-century industrialism. Similarly, Miller's observation that effective construal of kairos can "create opportunities for opportunities" points toward skilled rhetors' abilities to leverage situational awareness.

Richard Vatz responds to the deterministic assumptions of Bitzer's rhetorical situation and looks for an alternative characterization. Vatz adopts a contrary position that contends exigences themselves are fundamentally rhetorical, making them constructed realities that do not objectively exist. For Vatz, what matters most for rhetors is what he terms *salience*, the act of highlighting specific aspects of a context to create a perceivable situation (Vatz, 1973). Thus, rhetorical situations do not objectively exist to be responded to on their own terms but are rhetorically created from the beginning. All that is important to rhetors is their "decisions to make salient or not make salient" information in a manner that crafts a distinct situation (Vatz, 1973, pp. 156–157). Considering my grandfathers' argument, Vatz might contend that the real problem was not in their ability to produce arguments amenable to the situation but rather in their failure to characterize their concerns with enough rhetorical artistry to create an exigence salient to their interlocutor. Applying this viewpoint to Winner's description of high technology society would suggest failure on the part of technological critics to make salient the terms of their worldview and perhaps great success in the salience of technoutopian thought. Winner's arguments concerning *mythinformation*, or the widespread conviction that computer technologies will improve life, might be considered an instance of the computer industry making salient their goals (Winner, 1986, p. 105).

Like social constructionist theories of technology, the ability to speak an idea or object into existence is unlikely to account for the entire process. At best, such an ability is available for only those with the most control over discourse, while others are damned to a rhetorical hell of constraints. Bitzer's and Vatz's approaches might reasonably explain the dynamics of my grandfathers' argument; Miller's observations on the relationship between rhetoric and the theory of technological politics identify phenomena relevant to either interpretation. Like Bitzer's model, Vatz's model provides tools to interpret at least some situations concerning technology reasonably—but not all.

This antinomy is borne out when considering Winner's concerns about discourse in a high technology society. On the one hand, Bitzer offers an approach within which our tentative control of technological growth serves as an exigence for rhetorical action, inviting choice rhetorical strategies to address it that operate within constraints imposed by that context. Such constraints might include, but are not limited to, public belief in technoutopian rhetoric, government or corporate investment in the political affordances of given technological systems, technocratic inclinations, or even what Winner dubs *apraxia*—the tendency for a technical system to fall into chaos when "a significant link in the technical system ceases to function" (Winner, 1977, p. 186). On the other hand, Vatz might argue that rhetors should make salient less-than-savory facts and characterizations about technical systems they wish to address, pinning hopes upon rhetorically constructing a situation within which continuing certain ways of technological doing becomes untenable. Worthwhile salient facts might include, but are not limited to, environmental impact, record profits for those in charge of specific technical systems, marked changes in the standard of living for those who adopt specific technical systems, or even facts that suggest a technical system was explicitly designed with political purposes in mind. As Winner argues, only through public acknowledgment and deliberation of the realities of high technology society will we achieve "new spaces and roles for technological choice" that enable a world where "making things and taking action are one in the same" (Winner, 1992, p. 359).

12.2 Consigny: Rhetorical Topics as Synthesis

As suggested above, stakeholders or advocates engaged in discourse about technology would be foolish to adopt Bitzer's or Vatz's characterizations of rhetorical situations wholesale, lest they risk recreating the same contentious dichotomy between determinism and social constructionism that has played out in STS. Scott Consigny offers a conceptualization of the rhetorical situation that allows synthesis of Bitzer's situational exigence and Vatz's rhetorical salience, demonstrating how pragmatic rhetorical approaches offer a bridge between disparate philosophical commitments in technology studies and address concerns about discourse in technological societies raised by Winner. This synthesis suggests impactful rhetorical strategies for discourse about technology that do not solely root in deterministic or constructionist assumptions.

Consigny casts rhetoric as an art through which rhetors respond to extant situations by highlighting aspects of the context in unique ways. Central to Consigny's argument is the notion of rhetorical topics—set silos of assumptions from which enthymematic reasoning begins (Consigny, 1974). He argues that topics "structure context so as to open up and delimit a logical place in which the rhetor can discover and manage new meanings and relationships" (Consigny, 1974, 182). By understanding the situational nature of rhetoric as fundamentally related to topics, Consigny splits the difference between Bitzer and Vatz, requiring a certain amount of contextual guidance while also allowing space for rhetors to make salient choice terms and concepts. Consigny might address my grandfather's argument by noting how neither adequately addressed the topic "automotive design" because they failed to address the topic apart from their own opinions, making it impossible to craft a clear exigence or respond to common beliefs on the topic. In either case, Consigny's understanding of rhetorical situations provides adequate space within which we can both account for the logic that context foists upon rhetors and the artistic skills of those rhetors themselves.

Consigny's interpretation of the rhetorical situation is an inroad to reconsider Winner's theory of technological politics and Miller's rhetoric of technology.

Central to this consideration is Winner's characterization of the situation of modern technology. Again, Winner characterizes the modern technological context as one marked by an expansion of technical logic and erosion of public discourse concerning them, thus enabling the political nature of those logics to become influential. Winner does not look upon this situation and see technology directly determining our actions or unilaterally developing its own future. Rather, he identifies a variation of Frankenstein's problem, wherein humans create but forego much reflection on the implications of creation (Winner, 1977, p. 186). We liberally accept technological creation and conservatively critique it. Those who work toward democratic critiques of large technical systems are obliged to wrestle with this situation and the social beliefs from which it emerges.

Issues related to high technology society are interpretable as a distinct topic labeled "technological change" or "technological development." When considered by rhetors to leverage change upon large sociotechnical systems, this topic offers a distinct set of popular beliefs, political realities, bodies of knowledge, and legal frameworks with which rhetors must grapple as they establish meaningful relationships between ideas (Consigny, 1974, p. 182). As Winner argues, "so thorough is the penetration of technical novelty into American life that for a great many people, technology has become the very center of their self-understanding" (Winner, 1995). How this reality seeps into bodies of knowledge and political realities offers a definite set of constraints with which technologically minded rhetors must contend. Moreover, from a political standpoint, this formulation supports Winner's observation that modern technology creates a type of technological constitution, as "large sociotechnical organizations exercise power to control the social and political influences that ostensibly control them" and humans are presented with the need to evaluate "the material and social infrastructures specific technologies create for life's activities" (Winner, 1986, pp. 48–55). Responding to the exigence of eroded human control over technological systems, those inclined to activism must make salient "that a particular technical practice will generate some physically evident catastrophe" while simultaneously working from beliefs and ways of doing heavily influenced by the logic of technical systems themselves (Winner, 1986, p. 51).

Miller's conception of the rhetoric of technology overlays Winner's characterization of the modern technological situation and Cosigny's characterization of the situational nature of rhetoric: If Winner and Consigny synthesized suggest a concrete picture of rhetorical action concerning technology, then Miller identifies the dynamics therein. Importantly, Miller's tripartite conception of rhetoric and technology provides added complexity to Consigny's discourse-centered characterization of the rhetorical situation and Winner's more philosophically centered account of technology. Together, they form a complementary triangle, strengthening each other's analytic capabilities. Where Winner calls for citizen engagement and democratic deliberation, Consigny and Miller provide the tools through which action might be enacted. Where Miller identifies what might be considered rhetorical genres within the rhetoric of technology, Winner's influence ensures that she considers their material and political implications, and Consigny provides a situational framework within which they might be understood. Where Consigny offers a situational framework bereft of an applied context, Miller and Winner offer a distinct topic through which rhetorical constraints and rhetorical artistry can be enacted. If we are to address the very valid concerns raised by Winner concerning the political and seemingly autonomous nature of technological development in contemporary society, then addressing this topic with a clear understanding of rhetorical constraints and knowledge of the sources of rhetoric within the situation is vital.

Miller's conception of the rhetoric of technology also addresses a loose end that Bitzer, Vatz, and Consigny left unattended: material rhetoric. While the above formulation of the rhetorical situation in high technology society does well to address the modern complexities of rhetoric about and within technology, its rootedness in rhetorical scholarship from the 1960s and 1970s means that it does not consider materiality. Miller's work on the rhetoric of technology, which acknowledges a "pushmi-pullyu" dynamic between humans and technology, raises questions about how material rhetoric fits into our modern rhetorical situation. If, as Winner offers, technologies are a form of life, how does one contend with them in a rhetorical situation?

To my mind, this is the central difficulty encountered by those who engage in projects aimed at cultivating democratic sensibilities toward technology and the primary contribution of considering Winner's work in juxtaposition with rhetorical situations and the rhetoric of technology. Rhetoric about and within technology is intuitive because it centers on individuals and the human organizations they construct. When we advocate for technological change, we typically assume that our audience is humans or some organization they comprise, with the material nature of technology forming a type of ossified constraint. However, when we consider that the material affordances of a technology can produce a type of argument, another variable is added to the equation, for which any rhetor worth their salt must, in good conscience, account. This is not to suggest that humans can literally argue with technology. Instead, it is to suggest that when we engage with individuals who are thoroughly inculcated by technical logic, we are engaging with the technical logic itself. As both Miller and Winner remind us, the ethos of technology is often transferred to human rationality, meaning that any claim regarding technology produced therein might be considered a translation of sorts between humans and technical logic. Thus, what is most likely to become salient in any topic concerning technology is terms symmetrical with technical logic itself. The human players remain the same, but the rhetorical act and its situation take on the distinct timbre of technological society.

V

I wish to conclude by again returning to the recollection of my grandfathers. When I initially told their story, it was presented as if there were two main characters—my grandfathers themselves. But, this isn't entirely correct. We might identify the car itself as a third character, but I wish to push beyond this relatively obvious observation. Looking at the intersection of Winner's work with rhetoric, what seems to be the lurking variable is less the physicality of the car itself and more the technical logics that engender it. Remember again that my grandfather, who was a mechanic, was upset with my grandfather, who worked as a GM engineer to make vehicles increasingly complex. While this interaction might seem like the normality of two older men engaged in an argument, it points toward something more. Namely, my grandfather, who worked as a mechanic, intuitively understood my other grandfather as a stand-in for GM and for the decisions that the company made. My grandfather, who worked for GM, defended these decisions with the language of technical logic itself—for example, making things smaller and more compact was efficient—showing how his reasoning had become enmeshed with the sociotechnical system called GM. Both men wished to convince one another, and although they were not trained rhetoricians, they tried their damnedest. But, in accounting for each other, they only accounted for humanity. My grandfather, a mechanic, identified a clear exigence concerning technology, but he could not make his concerns salient because he offered them in the language of outrage and not in language symmetrical with technical logic. The topic was clear, but the audience was not. The dispute went unresolved.

While it is a colloquial example, my story and the juxtaposition of rhetoric with Winner's work offer the contours of a situation within high technology society. The technical systems that support our way of life continue to increase in scope and complexity. As Winner argues, our ability to comprehend the world we live in is increasingly eroded, and "the technical methods and symbols useful in creating an advanced technological environment are not those useful in making that environment intelligible to the mind" (Winner, 1975, p. 46). In this environment, we increasingly retreat to the comfort of technical logic, all of us safely sheltered within distinct silos of knowledge that do not ask us to think too far beyond our borders. This situation gums up our innate rhetorical ability to effect change, as it complicates the rhetorical situations within which we act. If democratic deliberation is a solution to the exigences presented by high technology society, then rhetors engaged within that rhetorical situation must learn to address the situation in a manner that makes salient identified problems *and* addresses those problems using language symmetrical with the technological society that has come to be. Only then will the (rightfully) incessant calls for effective citizen action on technological affairs make progress.

References

Barnett, S., & Boyle, C. (2016). Introduction: Rhetorical ontology, or, how to do things with things. In S. Barnett & C. Boyle (Eds.), *Rhetoric, through everyday things* (p. 3). University of Alabama Press.

Brummet, B. (1999). *Rhetoric of machine aesthetics* (p. 2). Praeger.

Consigny, S. (1974). Rhetoric and its situation. *Philosophy & Rhetoric, 7*(3), 181–185.

Hauser, G. A., & Benoit-Barne, C. (2002). Reflections on rhetoric, deliberative democracy, civil society, and trust. *Rhetoric & Public Affairs, 5*(2), 263–268.

Lyon, A. (2013). *Deliberative act: Democracy, rhetoric, and rights* (p. 18). Pennsylvania State University Press.

Miller, C. R. (1978). Technology as a form of consciousness: A study of contemporary ethos. *Central States Speech Journal, 29*(4), 234–235.

Miller, C. R. (1994). Opportunity, opportunism, and progress: Kairos in the rhetoric of technology. *Argumentation, 8*, 81–96.

Miller, C. R. (1998). Learning from history: World War II and the culture of high technology. *Journal of Business and Technical Communication, 12*(3), 301–302.

Miller, C. R. (2010). Foreword: Rhetoric, technology, and the pushmi-pullyu. In S. A. Selber (Ed.), *Rhetorics and technologies: New directions in writing and communication* (p. ix). University of South Carolina Press.

Sclove, R. E. (1995). *Democracy and technology*. Guilford Press.

Vatz, R. E. (1973). The myth of the rhetorical situation. *Philosophy & Rhetoric, 6*(3), 156–157.

Winner, L. (1975). Complexity and the limits of human understanding. In T. R. La Porte (Ed.), *Organized social complexity: Challenge to politics and society* (pp. 40–76). Princeton University Press.

Winner, L. (1977). *Autonomous technology: Technics-out-of-control as a theme in political thought* (p. 200). MIT Press.

Winner, L. (1986). *The whale and the reactor: A search for limits in an age of high technology* (p. 10). MIT Press.

Winner, L. (1992). Citizen virtues in a technological order. *Inquiry, 35*(3–4), 343.

Winner, L. (1993). Upon opening the black box and finding it empty: Social constructivism and the philosophy of technology. *Science, Technology, and Human Values, 18*(3), 368–371.

Winner, L. (1995). The enduring legacy of autonomous technique. *Bulletin of Science, Technology, and Society, 15*(2–3), 71.

Bitzer, L. (1968). "The Rhetorical Situation." *Philosophy & Rhetoric*, 1(1), 1–14.

Index

© The Editor(s) (if applicable) and The Author(s), under exclusive license to Springer Nature Switzerland AG 2026

A. Fox, D. Cressman (eds.), *Arrangements of Power: Tracing Langdon Winner's Legacy Within and Beyond the Philosophy of Technology*, Philosophy of Engineering and Technology 50, https://doi.org/10.1007/978-3-032-17371-3

MIX
Papier aus verantwortungsvollen Quellen
Paper from responsible sources
FSC® C105338

If you have any concerns about our products,
you can contact us on
ProductSafety@springernature.com

In case Publisher is established outside the EU,
the EU authorized representative is:
Springer Nature Customer Service Center GmbH
Europaplatz 3, 69115 Heidelberg, Germany

Printed by Libri Plureos GmbH
in Hamburg, Germany